Praise for *Planets for Pa*

"*Planets for Pagans* tr[illegible]s an enchanted journey. We learn about the stars and planets through engaging storytelling and cross-cultural myths amid the patterns and movement of the night sky. Renna Shesso has woven a luminous tapestry, sharing her insights, her personal experience, and her deep knowledge of ancient wisdom traditions. This brilliant book is a map that leads us back to our sacred selves."

—Sandra Ingerman, MA, author of *Soul Retrieval, Medicine for the Earth* and *How to Thrive in Changing Times*

"Wow, what a read! *Planets for Pagans* is an experience sure to intrigue your intellect, stretch your imagination, ignite your soul, and remind you of all the wonder you feel when you look up and connect with life above. As a nature lover and shamanic practitioner, I have always felt akin to the gems in the night sky, never feeling alone even while backpacking by myself in remote areas. Thanks to Renna, we now have the opportunity to understand more about these relations as presented through her masterful bridging of nature, astronomy, astrology, mythology, ancestral teachings, shamanic knowings, soul sensations, and simple truths. It's lovely for the head to understand what the heart has always felt, and the soul always known."

—Colleen Deatsman, author of *The Hollow Bone: A Field Guide to Shamanism* and *Seeing in the Dark: Claim Your Own Shamanic Power Now and in the Coming Age*

"At last! What a pleasure to have a book that looks at *our* sky instead of the abstract and dead universe of science. *Planets for Pagans* puts our forgotten, hidden sky into focus and brings its primeval meaning to us to provide food for our souls."

—Greg Stafford, author, publisher of *Shaman's Drum*

"*Planets for Pagans* presents a clear, accessible, up-to-date guide to the positions of the planets in the sky as well as to their resonances in our hearts and imaginations."

—Diane Wolkstein, co-author of *Inanna, Queen of Heaven and Earth*

PLANETS FOR PAGANS

PLANETS FOR PAGANS

SACRED SITES, ANCIENT LORE, AND MAGICAL STARGAZING

RENNA SHESSO

WEISERBOOKS
San Francisco, CA / Newburyport, MA

This edition first published in 2014 by Weiser Books, an imprint of
Red Wheel/Weiser, LLC
With offices at:
665 Third Street, Suite 400
San Francisco, CA 94107
www.redwheelweiser.com
Sign up for our newsletter and special offers by going to *www.redwheelweiser.com/newsletter*.

 Originally published as *A Magical Tour of the Night Sky* in 2011 by Weiser Books, ISBN: 978-1-57863-495-8

ISBN: 978-1-57863-573-3

Library of Congress Cataloging-in-Publication Data available upon request

Cover design by Jim Warner
Cover image: Constellation Card of Aquila and Surrounding Constellations
Library of Congress / Library of Congress
Typeset in Adobe Garamond and Frutiger
Interior by Dutton & Sherman

Printed in the United States of America
EBM
10 9 8 7 6 5 4 3 2 1

Contents

Acknowledgments

We come spinning out of nothingness, scattering stars like dust. Something opens our wings. Something makes boredom and hurt disappear. Someone fills the cup in front of us: We taste only sacredness.

–Rumi

With deep thanks and joyful gratitude:

To Jerry Davidson, Marilyn Megenity, Deb Hoffman, Cheryl Pershey, Rory Joyner, Jeanette Stanhaus, and Joy Vernon, my patient and outspoken readers, for their curiosity, encouragement, and late-night discussions. You helped keep the heart in this heady endeavor.

To the Colorado and wider communities of shamanism and the Craft, and the teachers, mentors, students, and friends I've been blessed to find within each. These connections with Spirit are my Polaris.

To the librarians of the Denver Public Library and the Denver Museum of Nature and Science's Alice Bailey Library; to Prospector, the interlibrary loan system; and to Starry Night Pro astronomy software, on which many of my illustrations are based. Your resources helped make this book possible (and any mistakes remain my own).

To Caroline Pincus, associate publisher at Red Wheel/Weiser Books and Conari Press, for your ongoing enthusiasm and support for this project.

In memoriam, to my wildly creative astrologer-grandmother, Freda Benson, and to pioneering authors Gerald Hawkins and John Mitchell.

And finally, to the backyards, dunes, and star-rich nights of childhood, where wonder began.

Chapter 0

Following the North Star

These are stories of love and Eternity, of deities and disguises, and of ancient knowledge that have receded into mystery. As mysteries go, however, this is one of the most playful:

The little girls are dressed as bear cubs, and they are dancing.

But let's start at the beginning.

First came wide-wandering Eurynome, Goddess of All Things, borne out of Chaos. Eurynome took the North Wind and crafted a great serpent from its air. She named this serpent Ophion; She danced with the snake and then She coupled with it. And then, from their mating, Eurynome birthed all things. All! Among the beings Eurynome brought forth were Phoebe, whose name means "bright," and Coeus, who ruled the intellect and the starry axis of the heavens. Together Phoebe and Coeus had a daughter, Leto.

Her name may come from *lethô*, "to move unseen." But Leto *was* seen. Zeus saw, desired, and courted Her. From Her mating with Zeus, Leto bore a pair of remarkable divine twins who, between them, governed Night and Day. They were called Apollo and Artemis—a god of the Sun and a goddess of the Moon.

Artemis was also the goddess of the hunt, the wild places, and all wild creatures, and was Herself devoted to independence. Four prancing, golden-horned deer pulled Her chariot. Sleek hounds and young mortal maidens were Her companions—among them Kallisto, "the fairest one."

Now Zeus saw Kallisto. He pursued her—was He disguised as His own daughter Artemis?—and caught her.

Kallisto remained among Artemis' companions until, one day when all were bathing, Artemis saw that Kallisto was with child. The goddess was furious. Pregnant women shouldn't masquerade as virgin devotees. So Artemis transformed Kallisto into a bear and, calling to Her huntress companions and Her hounds, set out to hunt Her former friend.

Zeus saw all this and intervened. He hid His lover, the bear-Kallisto, high up among the stars, along with a smaller bear—Kallisto's son, Arcus.[1]

One ancient ritual of Artemis was the *arteia*—"playing the bear"—during which little girls were dressed in honey-colored robes and yellow bearskins. Costumed as bear cubs, they danced to honor Artemis—as if in a mystery school's kindergarten play, but on a much larger, public scale that included entire city-states.[2] "Playing the bear" was a rite of passage for Athenian girls, their debut into the spiritual life of their city.

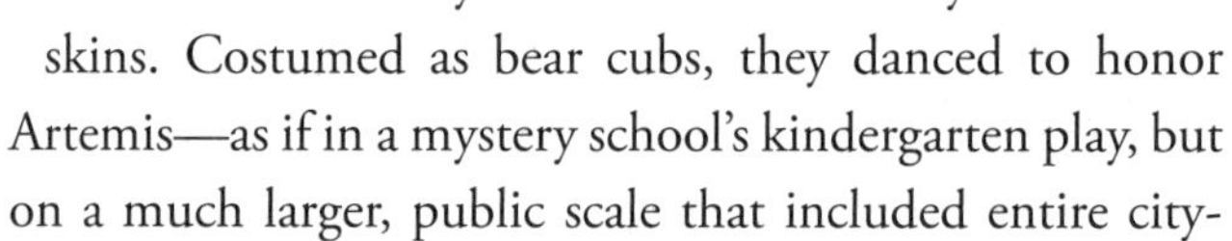

Figure 1. Ursa Major, the great She-Bear.

So, we begin our story with a goddess and her sensual dancing snake, beings who "move unseen," deities of Day and Night, shape-shifting gods, and the great night sky—especially the great She-Bear who still dwells among the stars, spinning nonstop above us (see figure 1).

The act of starting to write can be vertigo-inducing. In looking for my own starting point—and feeling as if I were going around in circles—I suddenly remembered that this is literally true.

I *am* going in circles. We all are. The Earth itself is rotating constantly, and we're all along for the ride. We observe this rotation when we watch the Sun "rising" and "setting"; but it's the Earth's own spin that accounts for the Sun's apparent journey across the sky.

Storytelling is sometimes called "spinning a yarn." It's a good image, and a recurring one. We weave a tale, picking up the thread of one theme or another as we go along. There are strands beyond number to the stories of the sky—all interwoven, perhaps even entangled—but most of them come back to that dizzying sense of the Earth's spin and how we and our very distant ancestors have tried to make sense out of the planets, the stars, and the Moon and the Sun whirling above us. This is a book of stories—of threads through time—spun, woven, and hopefully *un*tangled. We'll look at the lore of the heavens, some sky-related sites that our ancestors created, and ways of enlisting the sky as a spiritual ally *now*, in our own era. And—especially—we'll look at the sky itself.

As we do that, patterns will gradually begin to emerge, often against a background of deepest blue.

The Earth's motion is visible during the day by watching the Sun, but we see our planet's rotation more vividly in the night sky. Nowhere is it more obvious than around the "Pole Star," Polaris, also known as the North Star. That is our starting point. You can find the North Star using star patterns that many of us learn in childhood—the Big and Little Dippers, more formally known as portions of the Great Bear, Ursa Major, and the Little Bear, Ursa Minor. These two Dipper *asterisms* (recognizable star groups that aren't constellations) have long been known and valued throughout the Northern Hemisphere. The Big Dipper is the easier of the two to find, thanks to its size and brightness. Once you find the Big Dipper, you can find the North Star (the brightest in Ursa Minor) by using the pointer stars Merak ("the loin," or haunch, of the Bear) and Dubhe (Ursa Major's *alpha* or brightest star) at the edge of the Dipper farthest from the handle (see figure 2). This is wonderfully useful. Rather than trying to find Polaris, a single star, just search for the larger pattern of a recognizable group and proceed from there. Designations of *alpha, beta,* and other Greek letters were used early on to rank the relative brightness of individual stars within a constellation.

The Big Dipper is likely to be visible even when viewed from amid the competing lights of a city thanks to two things: Although it rotates, it stays put in the north, so we know where to look, and it has some stars bright enough to cut through city lights. We've become so accustomed to murky, not-really-dark city nights that many of us are startled by the visual overload of stars in a truly dark, rural sky. Ironically, on the other hand, urban light pollution makes it easy to spot the brightest stars by wiping out the competition, and the Big Dipper often manages to blaze through.

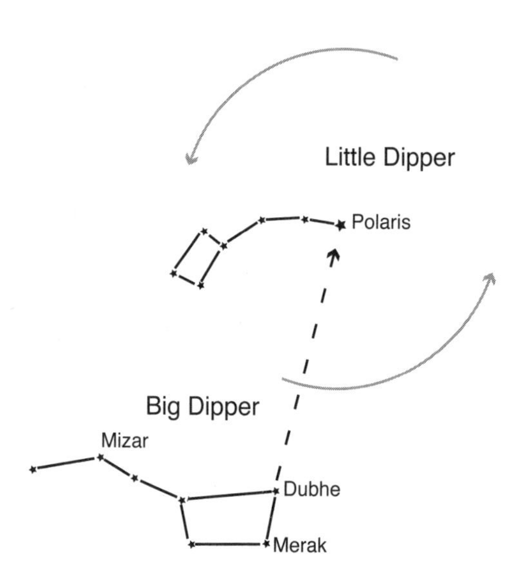

Figure 2. The North Star and the Dippers, with Merak and Dubhe as points on the Big Dipper.

Watch this area of sky some night for even half an hour or so and you'll see the steady counterclockwise motion that our Northern Hemisphere ancestors watched for millennia.

Above about 40° N (north) latitude, most of the Big Dipper is visible—or its pointer stars Dubhe and Merak, at least—every night of the year as it swings around the Pole Star like clock hands turning backward. It's circumpolar and thus never fully sets. As you get farther south, at about 20° N from the equator, the Little Dipper literally dips, as if scooping up water in its pan (see figure 3).

Latitude is measured horizontally, from the Earth's equator (which is 0°) to the North Pole and South Pole (which are at 90° N and 90° S, respectively). The higher the number of your location's latitude, the farther north or south you are. Longitude, the east-or-west vertical measurement, begins from a line running between the North and South Poles through Greenwich, England (site of the Royal Observatory, designated 0°) and moves east and west, measuring 180° in each direction, for a total of 360°. When used to measure the sky, degrees are applied as if projected forth against the inside of a vast sphere (astronomical) or as if marking locations along equal-sized zodiac signs (astrological).

Those of us who watch the Moon, the Sun, and the planets amid the zodiacal constellations are accustomed to looking in the other direction. We face south, focusing on a south-centered arc of sky where the planets and zodiac constellations are all located along the ecliptic, the Sun's path. But we have to face north to find the two Bears.

Watching the ecliptic, we see stars and planets move from left to right across the sky. That's how we read in English—from left to right—and that's how time passes on a clock, with the hands swinging up and over, left to right, clockwise.

But facing north, things are different. The stars move *counter*clockwise around Polaris. I sometimes feel a bit queasy when I watch the Bears, as if I were spinning in the wrong direction. Picture yourself standing along a one-way street filled with a parade heading west. If you watch from the north side of the street, you see them all moving from left to right; but if you cross the street and watch from the south side, you see them moving from right to left. The travelers haven't changed direction, but your perspective has shifted. Here's an additional consideration about which direction we face: When we look south, we see celestial motion as a huge arc between the eastern and western horizons, the visible portion of that rotating circle. Looking north, we see a full circle revolving around a central point.

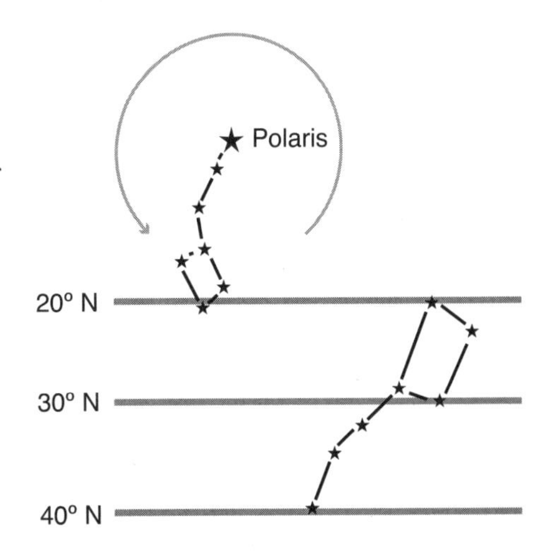

Figure 3. The Little Dipper "dipping." The Dipper's visibility shifts as we travel north or south.

We tend to equate "polar" with an earthly location—white bears, icy expanses, and intrepid explorers. But for early peoples, the Pole Star was perceived more literally as a pole—a stable center in the spinning sky, the picket stake around which everything else traveled, like a horse circling as it grazes at the end of its tether. The star at the center of all this action has been called Polaris since the Renaissance, the word coming from the Greek *polos,* meaning "pivot" or "axis." The movement of the other stars shows us the passage of time. The North Star, in its stillness, is outside of time and thus is a fitting symbol for Eternity. And indeed, to ancient peoples, it was the access point *into* Eternity.[3]

The North Star, Spinning, Norns, and Runes

How many of our tools and crafts gradually evolved out of this steady pirouette of motion? Polaris inspires human creativity, whether we tether a horse out to graze or spin a compass to create an arc on paper. We spin fibers, coil baskets and pottery, rotate our potter's wheels, "turn" wood against a blade to give it shape, and grind our grain between millstones. We replicate this perpetual spin in the twirling motion within our ancient and modern dances and in the spirals found in art from every era (see figures 4 and 5).

Figure 4. Stone spirals at the entrance to Newgrange.

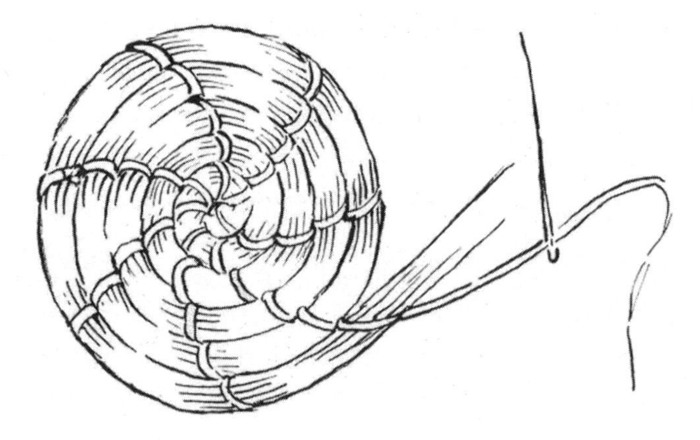

Figure 5. The base of a coiled basket.

Spinning is an ancient craft, and spun fiber fragments have been found that date back to Paleolithic times fifteen thousand years ago.[4] A single strand of sheep's wool or plant fiber has only its own strength; but when multiple strands are twisted together—spun—they become stronger, thicker, and far more durable. A good indicator of spinning's value is how well-populated our mythologies and folktales are with women who spin. One familiar example is the Moirae, the three Greek Fates: Clotho spun the thread of each life; Lachesis measured it; Atropos cut it. In the Roman version, these are the Parcae: Nona, Decima, and Morta. In the Norse pantheon, the Norns spun fates. In fact, the word *norn* may derive from a word meaning "to twine." The Norns' individual names—Urd, Verdandi, and Skuld—mean "past," "present," and "the results of the present"—a beautiful alternative to "outcome" or "future," as it leaves the way open for change. When we alter what we're doing *now,* we affect what will result later.

The Moirae are good examples of how characters and their tokens are used to express vast universal concepts. Clotho and her spindle and distaff represent birth, the creation of life. Atropos and her knife or shears represent death, as does the Grim Reaper with his scythe. Lachesis, with her measuring rod or ruler, seems to have the most passive job—measuring Clotho's thread. In fact, she represents life itself in its full duration, its length.

The drop spindle is a simple thing, a weighted stick at the end of a cord, similar to a plumb line or a pendulum except that the suspending cord is being created as the spindle twirls. The spindle's role is to keep the strands of fiber taut as they're brought together in a twist (see figure 6). Sounds easy, but it isn't. Spinning a fine, even thread requires skill, dexterity, concentration, and practice.

The spindle's weight, or whorl, is a fat disk, a conical bead, or a flattened sphere that is most often made of stone or wood. The whorl is shaped for good balance and drilled through with a hole to accommodate the spindle's stick (see figure 7). Large or small, ornamented or plain, spindle whorls have been found worldwide and are sometimes mistaken for beads. Spinning wheels were luxury items, but the

Figure 6. A woman spinning with a drop spindle and distaff.

Figure 7. Spindle whorls.

Figure 8. Drop spindles.

spindle and distaff (the long pole that held the un-spun fibers) were common tools among the less affluent.

In legend and lore, spindles have magical implications: Sleeping Beauty pricked her finger on a spindle's point and fell into an enchanted sleep; mounds of raw flax were spun in a single night; straw was spun into gold. Spinning may seem low-tech to us, but back when all tasks were done by hand, a whirling drop spindle was relatively automated (see figure 8). A bow-drill that made fire through friction was the next closest thing, both in automation and magical connotations.

Spinning was associated with women's concentration and craft, and with meditation and magic. Spinning with intention was a way to direct your energies magically—to manipulate reality like fibrous strands. A far-off battle might be spun into a husband's victory with help from his wife at home. Magical winds could be called forth by spinning, and sailors believed that spinning implements brought onboard ship provoked dangerous storms. The word *seidhr* refers to early Norse and Germanic shamanic trances and prophesying, but the word also connotes "string," "cord," "snare," and "halter." In some seidhr work, the intentions of the *völva,* the female shaman, are sent forth to *seidha til sin,* to "attract by seidhr," as if ensnaring something and pulling it to you, or manipulating an event from afar.[5] The völva's staff of office—her *seidhstafr*—looked like an ornamented distaff. We'll look more closely at distaffs in chapter 10.

Whether these simple tools were used magically or for mundane purposes, the spindle and the distaff were powerful, distinctly female implements. The wary tone of the Old European lore surrounding both spindle and distaff stemmed from the fact that their skilled use was so specifically the province of women and, by contrast, so thoroughly foreign to men. As seen in old burial goods worldwide, the spindle whorl often went with a woman to her grave.

The Scandinavian goddess Frigg was the wife of Odin and the patroness of marriage, childbirth, and the home. Domestic order—harmony within the home—is Frigg's purview, a vast realm of responsibility in an era when everything—cloth, food, furniture, fire—was produced manually from scratch. Folktales of young women who spin, or spin with the aid of magic, or refuse to spin at all, come from a time when all women acquired this skill and were well aware of its value—thread, cloth, clothing, trade goods, creative expression, prestige.

Besides its pragmatic use, however, spinning was persistently equated with women's magic. One of Frigg's attributes was the ability to see the fate of all beings, although She didn't reveal what She saw. Some tales credit Frigg with giving the flax plant to humans and teaching them how to spin and weave its fibers into linen.

The North Star was considered the point of Frigg's spindle, as if all creation were spinning forth from Her. This interpretation includes some distinct shifts, the first of which is perceptual. Where other stars are grouped to portray objects and beings, the North Star—taken on its own—can portray an object's *action.*

The second shift is one of perspective. The North Star's spin is perceived not as if we were watching someone else using a spindle but as if *we* are the spinner, looking down along the cord and the spindle's shaft to the creative motion at the end. When we envision Frigg out there in space wielding Her spindle with the thread emerging downward toward the Earth, it is we ourselves who are being spun into existence.

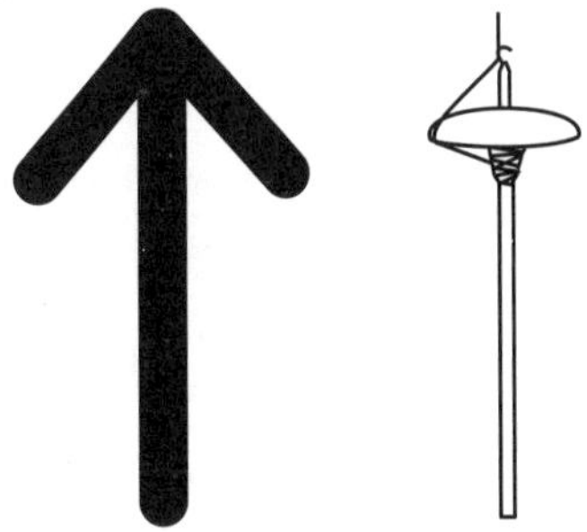

Figure 9. The rune Tiwaz and a drop spindle.

The rune *Tiwaz* is mainly identified with the law-and-warrior god Tyr (who we'll meet in chapter 7). The rune is even shaped like Tyr's spear—or like Irminsul, the world-tree of Tyr (see figure 9).[6] But Tiwaz has also been equated with Frigg's spindle—a top-whorl spindle, the type associated with Norse tradition.[7] This North Star/spindle interpretation gives Tiwaz an additional complex of meanings—creativity, choices, focus, direction—beyond its interpretive references to protection, justice, victory, and leadership.

One early Anglo-Saxon rune poem contains a commentary on Tiwaz that reads like a description of Polaris: "Tiw is a (guiding) star; well does it keep faith with princes; it is ever on its course over the mists of night and never fails."[8]

Time and Direction

Stars elsewhere in the sky change significantly throughout the year—some clearly visible in winter, others in summer. By contrast, Polaris and the Dippers are visible year round, although their orientation shifts constantly. Since the speed of rotation doesn't change, they are useful as a time-keeping device—like clock hands turning in reverse—and they have been used by many people for that purpose.

Looking at a single North American example, the Mescalero Apache of southern New Mexico watch the spin of *nahakus,* the Big Dipper, to gauge the pacing of nighttime ceremonies.[9] This is one of the roles of the singers—the tribe's doctor-priests, spiritual teachers and healers, medicine men, and shamans—who carefully pace their proceedings by the Sun and stars, so that all the necessary songs and rituals are accomplished in their proper way and time. *Nahakus* means both "that which revolves around a pivot point" and "falling," as the left-hand stars rotate downward around Polaris.

The north—as in that part of the sky, or that direction—has other meanings and uses as well. According to Mescalero Apache singer Bernard Second, this direction symbolically encompasses his people's ability to forge through life's difficulties, the cleansing powers of the snow, and The Land of Ever Summer, the ancestral home to which they return at death. In burial, a body is oriented with the head to the north, and the first in the ring of stones encircling the burial mound is placed at the north.

The directions taken together are symbolized by the quartered circle (see figure 10), which is now commonly called a "medicine wheel" (not a term Second used). The medicine wheel signifies everything: the universe, the sacred number four, harmonious and balanced life, time. Second's student, ethnographer Claire Farrer, describes this "base metaphor" as expressing two kinds of time—the vertical line as the Sun's yearly north-to-south shift, and the horizontal line as the Sun's daily east-to-west journey across the sky. This symbolic shape underpins all Mescalero Apache activity: how they speak (a good presentation has four parts), how they sprinkle salt on food (a single pinch of loose salt, sprinkled in a clockwise circle, then in two crossing lines), and their four-armed-star basketry design. Any action done properly helps create *h*^{n}*zhúne,* which encompasses goodness, beauty, harmony, balance, health, and well-being.

Figure 10. The quartered circle, often called a "medicine wheel."

In Mescalero Apache terms, to spin yarn counterclockwise—as we see the Big Dipper rotate—is to invite sickness and disorder.[10] In their basketry, however, either direction is acceptable, although a basket woven counterclockwise—in the direction of the Dippers' movement—is limited to everyday use, unless blessed by a singer. Baskets meant for ceremonial use are woven clockwise, which is awkward and far more challenging for anyone who is right-handed.[11]

2° wide

Figure 11. Your index finger as a measure of 2° against the arc of the night sky.

Navigation—The North Star and Travelers

Eternal and consistent as the Earth and star motion appear, the star designated as the North or Pole Star changes over time, as the Earth shifts gradually on its axis. Ursa Major and Ursa Minor are always near the center of the northern action, but the Little Bear's tail star has only been considered the Pole Star since about 1100 CE. Polaris' North Star position isn't completely exact even now, although it's less than 1° off true celestial north. When you hold up your index finger to the night sky, its width is about 2° (see figure 11). Half of that distance, a single degree of deviation in space, won't cramp the style of any determined travelers. We'll look at this shift in chapter 3.

As long as they could recognize a pattern of stars that allowed them to find north, our ancestors could extrapolate the other directions. Earlier travelers employed all available resources, not only the North Star, to set their course. Geological points on land (landmarks) and a variety of other recognizably consistent characteristics—wave patterns between remote islands, for example—have allowed humans to use all their ingenuity, all their senses, to move from place to place in their explorations. The North Star, while important, was thus just one tool among many.

North Star navigation works like this. As you travel south, the North Star gets ever-lower in the sky; when you go north, it gets higher. If you travel from east to west, you can stay on a relatively direct line by checking that the North Star is at a consistent level off your right shoulder.

Ironically, the Pole Star is least useful for travelers in the far north, especially near the North Pole: this is because it is so directly overhead that they have difficulty gauging its relation to the horizon. Moreover, like all the stars, it's invisible during the summer months when the polar Sun doesn't set: no night darkness, no star navigation.

The Road to Freedom

Prior to the Civil War, escaping slaves from America's southern states made their way to the free states and Canada by following the North Star or the more recognizable Dippers, which they called the Drinking Gourds.

There's plenty of history—written and oral, songs and lore—about escaping slaves and the North Star. Millions of slaves were held in the South. In 1860, slaves accounted for nearly one third of the twelve million people in the fifteen slave-holding states. Those determined or desperate enough to risk escape used every means possible to share the information needed to make a successful break for the North. Written messages were rarely an option, as few slaves were literate. Although some progressive owners quietly taught individual slaves how to read and write, it was illegal to do so. Other means of communicating information in secret were necessary.

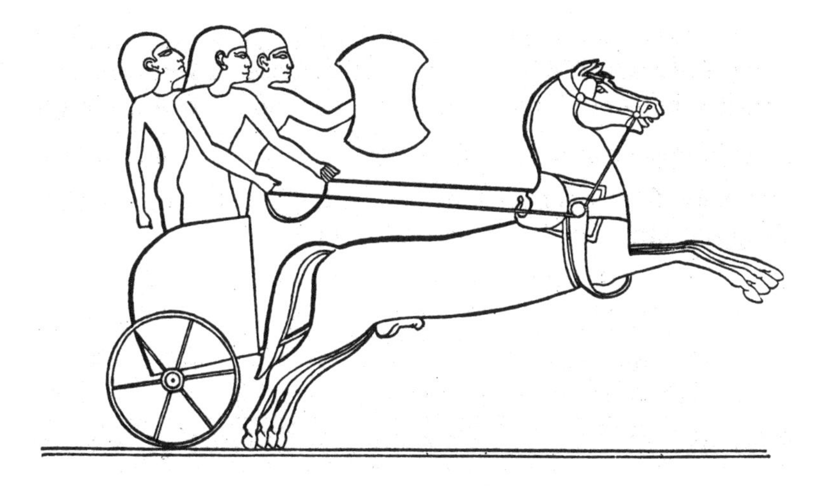

Figure 12. An Egyptian chariot. The shape of most simple two- or four-wheeled carts is a good match for either Dipper.

While some of those who escaped to freedom as independent travelers went on to tell their own stories—Harriet Tubman and Frederick Douglass are key examples—many written accounts of escape come from Underground Railroad sources. And many of these histories mention North Star–related songs.

When the old chariot comes
I'm going to leave you
I'm bound for the promised land...

Though overtly religious, the song "Old Chariot" was sung by Harriet Tubman as "a public proclamation" to let people know that they should get ready to leave.[12] The chariot image is biblical, but Big Chariot and Little Chariot are also old folk names for the Dippers. Thus the song is also a reference to the North Star and escape, with the "promised land" being, not Heaven, but the slavery-free Northern states. Another song, "Swing Low, Sweet Chariot," used the same imagery, and its phrase "Coming for to carry me home" was a practical reference to the chariots (or Dippers) moving around Polaris in the night sky. One of Tubman's code-names was "Old Chariot"—rhyming with her own name, Harriet—which makes this remarkable woman synonymous with the guiding North Star.

When the sun comes back and the firs' quail calls
Foller the Drinkin' Gou'd...
For the old man say,
"Foller the Drinkin' Gou'd"... [13]

And tens of thousands *did* follow the drinking gourd north to freedom.

Quilts were also used as visual signals to pass along information. According to Ozella McDaniel Williams, an African American quilter, a code involving numerous quilt patterns was used in the pre–Civil War years to signal slaves to prepare for escape, and to detail strategies and routes.[14] When hung outside to air, a large quilt was visible from a distance, like a flag. When stitched up as single "blocks"—the way seamstresses shared quilt patterns in pre-magazine times—the designs could be passed innocently from hand to hand. Traditional patterns like "Bear's Paw," "Compass Point," and "North Star" each had obvious associations with Ursa Major, Ursa Minor, and Polaris (see figure 13).

Figure 13. Top to bottom: Bear's Paw, Compass Point, and North Star quilt blocks.

Thus the North Star was a navigation tool and a symbol of hope. As he rose to prominence as an abolitionist orator, the brilliant former slave Frederick Douglass founded a newspaper. Begun in 1847, the paper's motto was "Right is of no Sex, Truth is of no Color, God is the Father of us all, and we are all brethren." Douglass named his newspaper *The North Star.*

Star Travelers—Chinese Shamans and the North Star

Taoist mystics had their own means of Great Bear–inspired internal "navigation." Working in trance, they mentally traced and retraced the patterns between the constellation's stars until they moved among their inner stars easily, far outside of time and space. This was travel in the sense of shamanic journeying—the journey of the soul—and was called "pacing the void."

Shamans-in-training had a more grounded version of this practice called "pacing the Yu." As if moving through the steps of a dance or a Tai Chi Chuan figure, they closed their eyes and physically stepped along the envisioned pathways between the Big Dipper's stars. In another variation, travelers lay down on a large diagram of the asterism and moved along its pathways in their mind's eye, using prayers and patterns of breathing as they did so to bring them ever closer to the "solar luminosity" of the star Dubhe.[15]

Spindle whorls have a look-alike cousin in the Far East, as they bear an uncanny resemblance to the ancient Chinese *bi,* jade disks that have been meticulously crafted and highly valued for roughly five thousand to six thousand years. Research holds that *bi* were never used as spindle whorls, but their shape is nearly identical to the flat-doughnut style of the whorl shown on the left in figure 7. Moreover, like whorls, the *bi* vary from small to large, and from plain to highly carved. *Bi* were presented as tribute and kept on altars. Many have been found among burial goods in the graves of men, where they had the special function of guiding the departed spirit into the heavens and acting as a portal for communication with the ancestors. And it is of particular interest to us here that the hole in the *bi*'s center was symbolic of the circumpolar stars—fixed and eternal, the place of immortality, the Absolute—around which all else spun.[16]

Early Great Bear Myths

In the *Odyssey*, Homer mentions "Wagon" or "Wain" as another name for the Great Bear, which never sets—"never dipping into the stream of Oceanus." Ulysses navigates by these stars as he sails away from Calypso's island.[17] To the Babylonians, the Big and Little Dippers were the Wagon and the Wagon of Heaven respectively. The goddess Ninlil, Lady of the Winds, was the ruler of these stars, which were used magically for prophetic dream work, especially to gain celestial permission when considering a trip. Supplicants went up onto their home's flat roof—a popular sleeping place during hot weather—and sprinkled libations of water toward the Wagon three times, then dusted a handful of flour over the rooftop. That night's dream would bring signs of the future, and if the supplicants dreamed of receiving a gift, their travels would be a success. Physicians also checked with the Wagon for indicators on the fate of their patients, since the terminally ill couldn't die without the Wagon's "permission."[18]

In Welsh lore, the Great Bear is King Arthur, *arth* being the Welsh word for "bear." The once and future king, who will return when needed, is the never-fully-setting Bear, who remains constant until that time of need.

Figure 14. Tarot Major Arcana card VII, the Chariot.

The Dippers, the Number Seven, and Tarot

While the Ursa constellations are consistently identified as bears throughout the Northern Hemisphere, the Dipper asterisms have a range of folk names. Chariot, Plough, Dipper, Wagon, and Wain are all used across a wide range of eras and regions. *Wain* is an old term for "chariot." As a simple two- or four-wheeled cart, its shape is a good match for either Dipper (see figure 12). Each of the Dippers has seven bright stars, and that number often appears as part of these asterisms' names—for instance, *Tseih Sing,* "the seven stars," for the Chinese and *Sapta Rishi,* "the seven great Sages," in Hindu tradition.

We use the Big Dipper to find Polaris, and then use Polaris to find our direction, to set our course. In the same way, setting or clarifying our course through life is one interpretation of the Chariot card in the tarot, where it is number seven—VII—among the Major Arcana cards (see figure 14). Tarot author Oswald Wirth associates this card with Ursa Major.[19]

Viewing the Dippers and the North Star

Figures 15 through 18 show the northern sky with the Dippers as seen at midnight through the seasons.[20] Using these four equidistant points in the year, we see fairly precise quarter-turns in this wheel. The Dippers each make a full circuit around Polaris each day. Watch them in any season and you'll see them rotate *counter*clockwise.[21]

Mizar, the second star from the end of the Big Dipper's handle (see figure 2), has a much fainter companion star called Alcor. The pair were known as the "Horse and Rider" in Arabia. The Roman armies used them as a vision test for

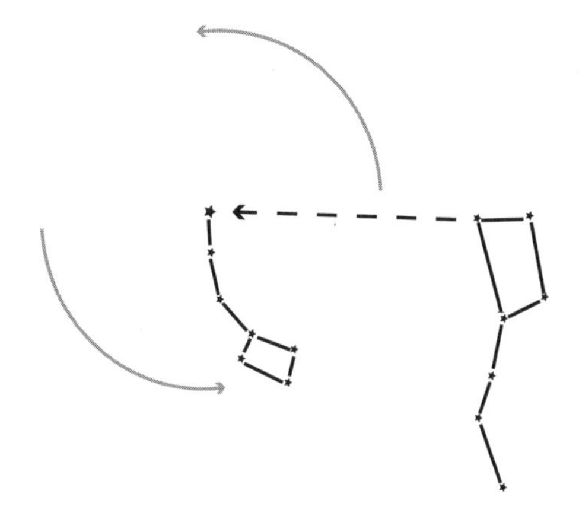

Figure 15. Winter Solstice at midnight.

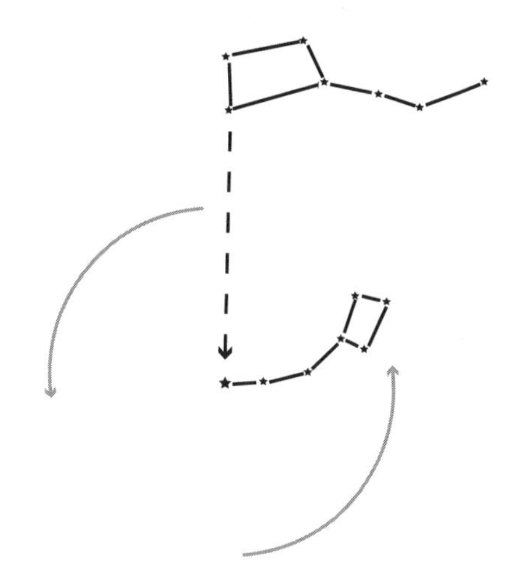

Figure 16. Spring Equinox at midnight.

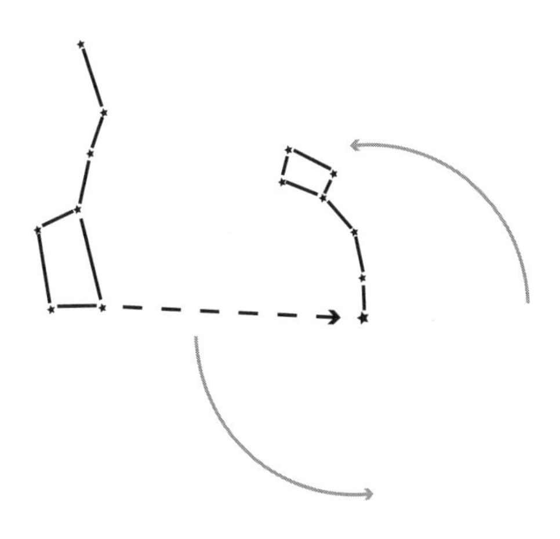

Figure 17. Summer Solstice at midnight.

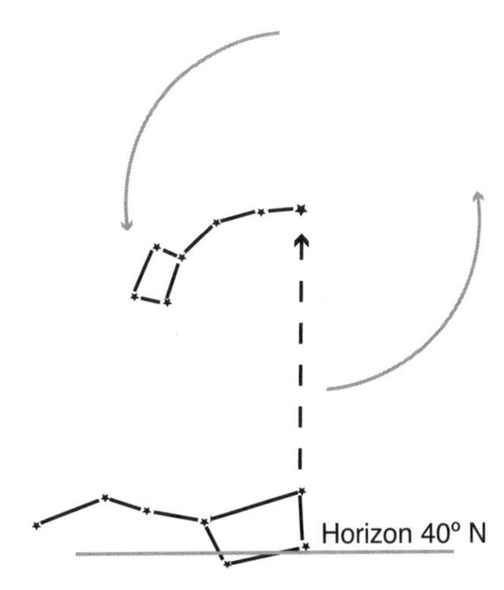

Figure 18. Autumn Equinox at midnight.

prospective archers—if you couldn't see Alcor as distinct from Mizar, no job for you. For the Babylonians, the Dipper asterism was a Wagon, with Alcor as a small fox perched on the wagon's tongue—trickster fox as trickster star.[22] Other tales identify Alcor as Odin's "night raven"—Nachtrabe—or as Odin himself transfiguring into that bird.[23]

So this is our starting point, the North Star, and off we go from there. The sky was our original calendar, our original storybook, the first illustrated edition, the prototype GPS. Beyond its pragmatic usefulness, the sky was the domain of spirit, traversed by deities and a place to which human souls departed. Let's re-enchant it, shall we?

Learn by Doing

Reading and studying are great, but physically *doing* something embodies knowledge within us at a much more visceral level. The activities at the end of each chapter in this book will help to plant some of these seeds. Engage your imagination and create your own activities, but be sure to try the ones mentioned below, as they will help you get your sense of direction tuned up. From here, we'll head out for the rest of the sky.

1. **Get to know the North Star:** Find the Great Chariot and Polaris in the night sky. Note their positions relative to each other. Watch this shift over the course of an hour, a few hours, over the next month, over an entire year. When outside at night, glance to the north and spot the Dippers to get your bearings. Does doing this shift your sense of orientation? Especially if you're accustomed to looking east, to south, to west for the Moon and planets, does the counterclockwise northern spin around Polaris affect you in any way? How?

2. **Consider a basic spinning class:** The act of spinning is mesmerizing and can be a form of meditation through focus. It's also tactile and connects you back thousands of years to the roots of human ingenuity.

3. **Learn something about bears:** Visit a zoo, or get some videos or books from your library. There's a wealth of material on this animal—from their real-life nature to their wide representation in folklore.

4. **Explore lucky number 7:** Seven is considered a lucky or important number in many traditions. What personal associations does it have for you?

5. **Try some tarot-related work:** Start with card VII, the Chariot. Using the image on the card as your starting point, "enter" the card. Visualize yourself as the charioteer. Check your orientation. Are you aiming toward the North Star to explore its symbolic meanings? Or are you heading away from it, setting out on a personal quest? If you feel stuck, or if you haven't liked the results of some recent choices, explore decision making as an overall theme. If you own several tarot decks, lay out several Chariot cards and spend some time exploring their similarities and differences, both for what the cards contain artistically and for what each may evoke for you.

6. **Try some North Star rune-related work:** Start by using Tiwaz. Selecting intuitively, let yourself be drawn to a tree. Sit at its base on the tree's south side, so the tree is at your back to the north. For the moment, let this be your World Tree, your supportive center. Use this as the starting point for a meditation, a shamanic journey, or some free-association sketching or journaling on whatever arises.

Chapter 1

The Zodiac—Our Circle of Animals

Are the constellations the first human artworks? We didn't create the stars, but we learned to recognize the patterns they form, connecting the dots to extrapolate specific figures from a bewildering ocean of individual sparks, imaginatively filling in the details that flesh out simple stick figures.

By recognizing patterns in the stars, we remember our ancestors and their stories about how our world was created. In this way, the sky was our original storybook, map, art gallery, and reference library.

When we look at the skies now and scoff—"Well, that sure doesn't look like a hunter (or a bull or a goat-fish) to *me*"—we're missing the larger picture, in every sense. Pattern recognition is a considerable survival skill. The patterns inherent in flowers and leaves help us separate the edible plants from the poisonous ones. The normal growth patterns of foliage help us notice when the patterns are amiss: Not *just* shrub! Shrub with hiding lion! Not *just* tree branch! Snake on branch! Pattern lessons were important in making split-second life-and-death decisions, and we're here today because our ancestors learned those lessons. So let's not denigrate those survival skills. Who knows if our minds still have the imaginative agility and rock-steady memory power to match those achievements?

One of the earliest patterns to catch our ancestors' attention was the motion of the Sun and the Moon. Although the Moon wanders widely from side to side, the Sun keeps faithfully to its path. The trail markers along the Sun's route are particular groups of stars positioned end to end in a great circular band, like a celestial storyboard. The width of this path is considered to be 20°, measured as 10° to each side of the Sun's own route, the ecliptic. You can roughly measure 20° of sky-space using your hand. Extend your arm, then spread your thumb and pinkie to their full extent (see figure 19). Keep the rest of your fingers up if you prefer; I tuck mine out of the way.

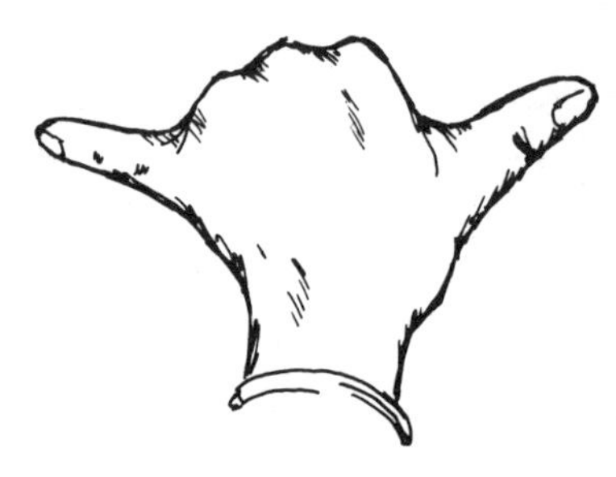

Figure 19. A hand measuring a 20° span of space in the sky.

First, get oriented. Stand facing south with your arms stretched toward the horizon on each side. Picture the Sun rising up to meet your left hand, arcing up and overhead during the day, and then sinking into the west past your right hand. Below the horizon line, unseen, the Sun continues its journey, as if moving underfoot. With dawn, it reappears again at your left hand, perpetually inscribing a giant clockwise—"sunwise" or "deosil"—circle. Its route is the ecliptic, the Sun's path through the sky. Yes, we're actually the ones moving, as the Earth orbits the stable Sun, but we need to describe what we see. "Earth turns Her back on the Sun to make darkness" is a much more poetic and accurate description of this phenomenon, but "sunset" is simpler.

The Sun never comes anywhere near most of the eighty-eight official constellations in the sky, moving instead in front of a very specific selection of star groups. Just like Earth, the other planets also orbit the Sun, so they also travel along this narrow path of specific constellations. Over the millennia, our ancestors codified their means of describing the Sun's journey by using these Sun-path constellations, even though different civilizations perceived the star groups differently (i.e., dippers versus chariots). Many traditions recognize twelve ecliptical constellations, but that number can vary from culture to culture. Most of ours represent animals, real and mythic, so they became known as the zodiac, from *zodiakos kuklos,* Greek for "animal circle."

We use this animal circle as a kind of map to illustrate the narrow portion of the sky shown in an astrological chart, useful for orienting ourselves to the sky. The zodiacal constellations become a twelve-slice pie, generally oriented with Aries just *below* the left-hand eastern horizon and Libra just *above* the right-hand western horizon (see figure 20). The circle's top is "up," not "north," as on a road map. The upper half signifies the visible sky, above the horizon. The half-circle arcing below is the non-visible sky, below the horizon on the other side of the Earth. In reality, this circle turns clockwise, as different constellations take turns being at your east-pointing left hand.

Around 2400 BCE, the Babylonians developed celestial and mathematical concepts based on dividing a circle into 360° relating to early ideas about the number of days in a year.[24] Our calculation of a year's length has improved, but we still divide circles into 360º. For *astrological* purposes, each of the "signs"—the zodiacal/astrological constellations that mark the Sun's path—is allotted an equal 30º. Like the numbering in the tarot's twenty-two Major Arcana cards—0 through 21—the degrees are numbered 0º through 29º.

So far this is an astro*logical* perception of the sky, however. Astro*logy* views the sky through the magnificent lens of the human imagination, with all the creativity and metaphoric symbolism of the ages added to the mix, while in fact our real goal here is to get back to the roots of these stories—ideally by stepping outside and seeing the original illustrations, which is astro*nomy*.

Figure 20. The astrological circle.

Our vague, in-the-eye-of-the-beholder boundaries between constellations were officially reinterpreted in 1930 when Belgian astronomer Eugene Delporte redrew the boundaries on behalf of the International Astronomical Union (IAU), straightening out earlier curved and diagonal interpretations of these nonexistent boundaries so they could be expressed more precisely, as shown in the illustrations in this book.

Even before Delporte revamped the boundaries, however, another important factor was in play. The constellations of the zodiac actually occupy varying amounts of space along the ecliptic, *not* identical slices of exactly 30°.

Star Motion

We just talked about circumpolar stars—those that never set. As we look to the zodiac and the planets, we are dealing with celestial bodies that rise and set—that is, they appear in the east and then disappear in the west. Unlike the circumpolar stars, not every star is visible all year long.

Measured against the backdrop of the zodiacal constellations, the Sun shifts eastward in the sky at the rate of about 1° a day, so each star gradually disappears from view. One day we see a given star, briefly visible low in the western sky just after the Sun goes down. The next day, we don't see it at all, because the Sun has moved another 1° eastward, bringing it too close to the star we're watching for us to see it. The last day a star is visible in the west at sundown is called its *heliacal setting*, from Helios, the Greek Sun god.

Only the brightest stars display this heliacal activity, since they compete with the brightness of the Sun itself, with only the onset of twilight in their favor. The dimmer stars set invisibly between daylight and twilight, but the brighter stars are visible to us in this quick-blink heliacal setting once each year. There's no firm mark-your-calendar date for this moment. Over the generations, however, people developed good knowledge of this based on their own geographic location.

Anything directly in line with the Sun won't be visible for a while. That's called a conjunction—when two planets, or a planet and star, are in line with each other and thereby visually conjoined.

So the Sun moves eastward about 1° a day. As it does, stars gradually reappear just before sunrise in the eastern sky. One day, a given star isn't visible because it's directly in line with the rising Sun and lost in the glare. But a few days later, there's our star—a blink of light!—visible for an instant in the pre-dawn twilight before the rising Sun quenches it. This first reappearance is called the *heliacal rising*. The next day the star is visible for a moment longer, getting farther ahead of the rising Sun. As weeks pass, the star and its constellation gradually rise long before the Sun, and are visible throughout the night.

With dedicated sky-watching and a stunning awareness of small details and subtle changes, our ancestors paid great heed to heliacal risings, these first-of-the-season sightings. Remember, heliacal risings and settings are momentary occurrences, not leisurely night-long events. Their actual dates are consistent to within two or three days,[25] but that's influenced by visual acuity, weather conditions, and vantage point. Only through acute observation can these occasions be witnessed. And when a star reappeared, it was as if all the energies and associations of that star were symbolically reawakened and reborn.

We'll look more at some heliacal risings in chapter 10. For now, just know that this is how the Sun shifts in relation to the zodiac. When the Sun occupies a constellation is when we *can't* see that group of stars.

Finding the Zodiacal Constellations in the Sky

The zodiacal constellations help us get our bearings on the rest of the sky and deepen our comprehension of the seasons and of the universe through which we spin. In figures 21 through 33, stars marked by the largest star symbols can often be seen even in a large city's light-polluted sky. Once you learn to recognize their relation to each other, you'll be able to distinguish them earlier in the evening, as they emerge after sundown.

At midnight on the Winter Solstice, you can find the first three signs of the zodiac high overhead. To locate them, find Orion's Belt. Follow its line to the right and up slightly for just over 20° (one hand span) to Aldebaran, Taurus' brightest star and the charging Bull's right eye (see figure 22). *Aldebaran* means "the follower," since this star follows the noteworthy Pleiades cluster. To find this

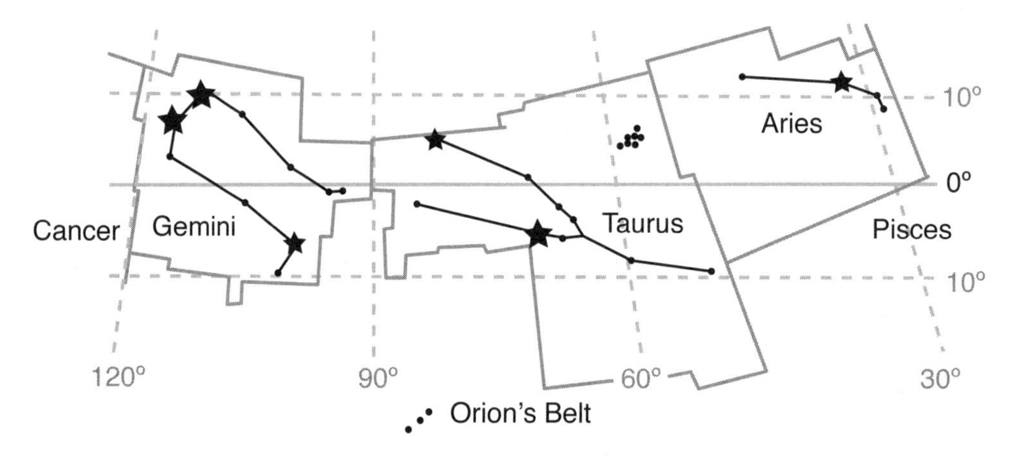

Figure 21. Aries, Taurus, and Gemini (right to left) as seen in the winter sky around midnight on December 20, or around 9 p.m. on February 1. They'll be high in the sky. The solid gray horizontal line represents the ecliptic.

group, continue to the right and up, to what may look like a small pale smudge. This is the Pleiades. From Aldebaran, go up and left to Elnath, which is brighter than the Pleiades but less distinctive and less bright than Aldebaran. Elnath, the Butting One, is the tip of the Bull's left horn. Note that, in summer, the Sun's path—the ecliptic—is high overhead in the daytime and low toward the southern horizon at night. The reverse is true in the winter. That's because the Earth sits at a tilt. The ecliptical stars visible during the winter months are those riding the higher side of our tilt. This makes for easier and more dramatic viewing of the sky in the winter months.

Returning to the Pleiades, go right 20° to find Hamal, Aries' brightest star, from *Al Ras al Hamal*, the Head of the Sheep (see figure 22). To Hamal's right is Sheratan, called the Sign because of its old role as the beginning of the zodiacal year, when its rising marked the Spring Equinox.

Returning to Orion's Belt, look up and to the left, a little more than 20° away, to see Gemini's fairly bright Alhena (see figure 23). Another 20° from Alhena, farther up and to the left, are Gemini's "twins," Pollux (south) and Castor (north and slightly brighter). These two are 4.5° apart—a bit more than two finger widths—and, like twins, are so well matched that they quickly become familiar.

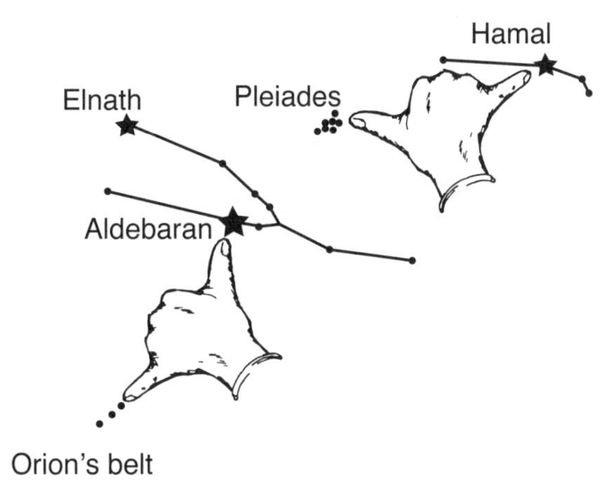

Figure 22. Finding the stars Aldebaran and Hamal.

On the Spring Equinox, a different zodiac group occupies the midnight sky (see figure 24). Find Leo by using the Big Dipper's pointer stars in reverse. In roughly 40°—two hand spans—you'll be in the mid-zone of Leo (see figure 25). Leo's brightest star, Regulus, is on the right and sits almost exactly on the ecliptic. Regulus ruled the heavens—it was the Babylonian's King Star—as the Lion's heart. The Babylonians anticipated eclipses when the Moon moved across Regulus, since this meant that the Moon was directly along the Sun's path. Among Leo's dimmer stars, Algieba is above Regulus as the Lion's mane; 24° to the left is Denebola, from *Al Dhanab al Asad*, meaning "Lion's tail." Leo actually looks like a lion, seated in a regal, sphinx-like pose.

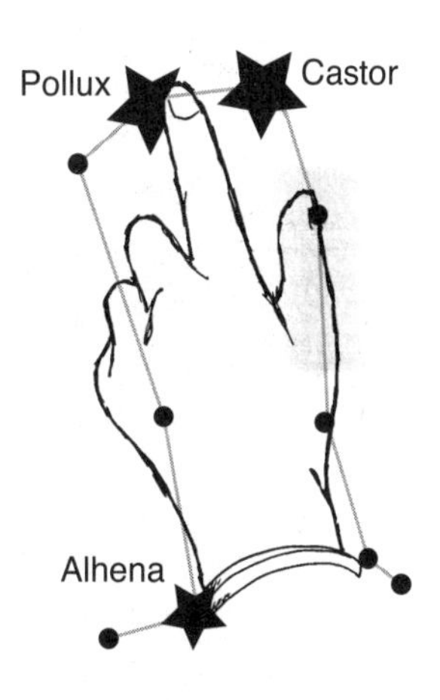

Figure 23. Finding the stars Castor and Pollux.

Cancer lacks bright stars, so find Gemini's Castor and Pollux, and then find Leo's Regulus. What's in between them is mostly Cancer. Procyon, the bright star in Canis Minor ("the Little Dog"), is another good landmark below Cancer (see figure 26).

To Leo's left is Virgo, the Maiden, occupying the widest space along the ecliptic, about 44°. To find Virgo, look north to the Big Dipper's handle. Follow its curve out in an arc, away from the ladle, to the first

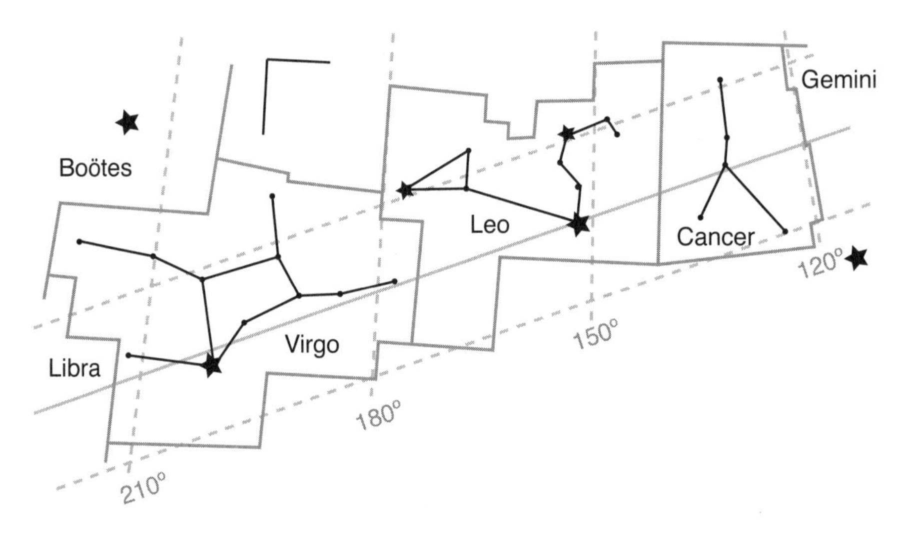

Figure 24. Cancer, Leo, and Virgo (right to left), as seen around midnight on March 20, or around 9 p.m. on May 1.

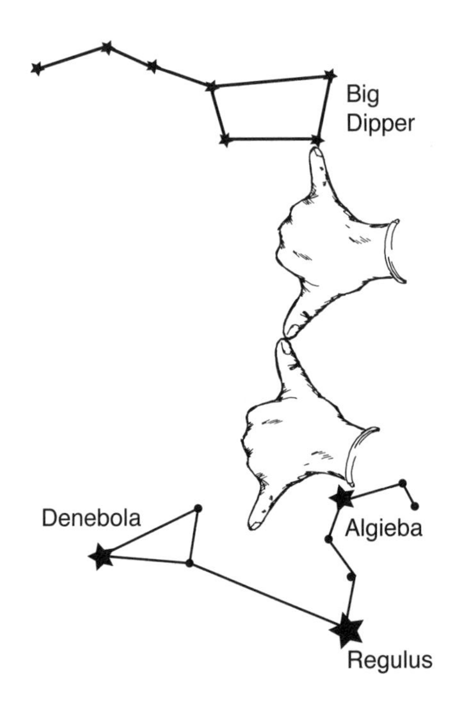

Figure 25. Finding Leo using the Big Dipper.

really bright star, Arcturus, in Boötes about 30°, or one-and-a-half hand spans, away. Continue this arc for a little more than 30° to the next really bright star, Spica, Virgo's brightest star, the ear of wheat held by this grain goddess (see figure 27).

Midsummer at midnight, the next group of stars rides low along the southern horizon. The easiest star to spot in this group is reddish Antares, Scorpio's brightest. It is named for the red planet Mars—*ant-ares* means "similar to Ares"—and is considered the Scorpion's heart. Scorpio is long from north to south, but also narrow, officially occupying only about 6° of space along the ecliptic, the least of any of the twelve signs of the zodiac (see figure 28). As with Leo, Scorpio looks like its namesake. The hooked tail is unmistakable, a tall scorpion with its tail dangling low.

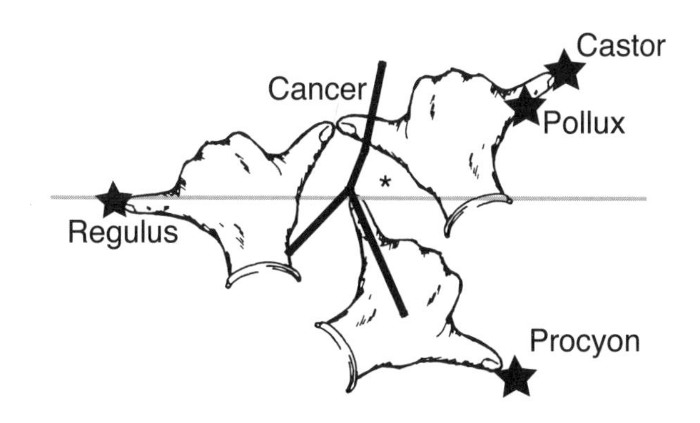

Figure 26. Finding Cancer.

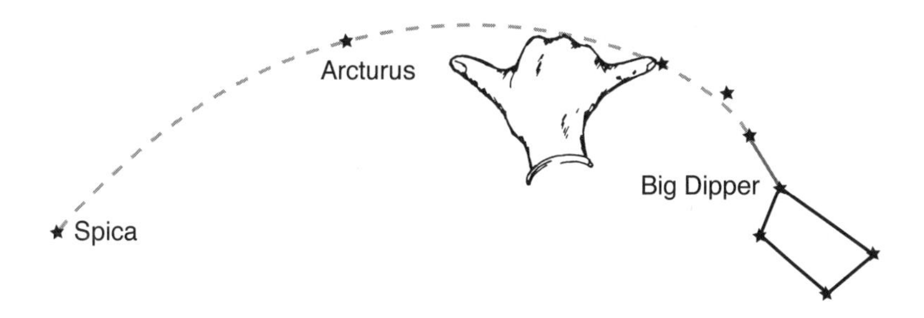

Figure 27. Finding Spica.

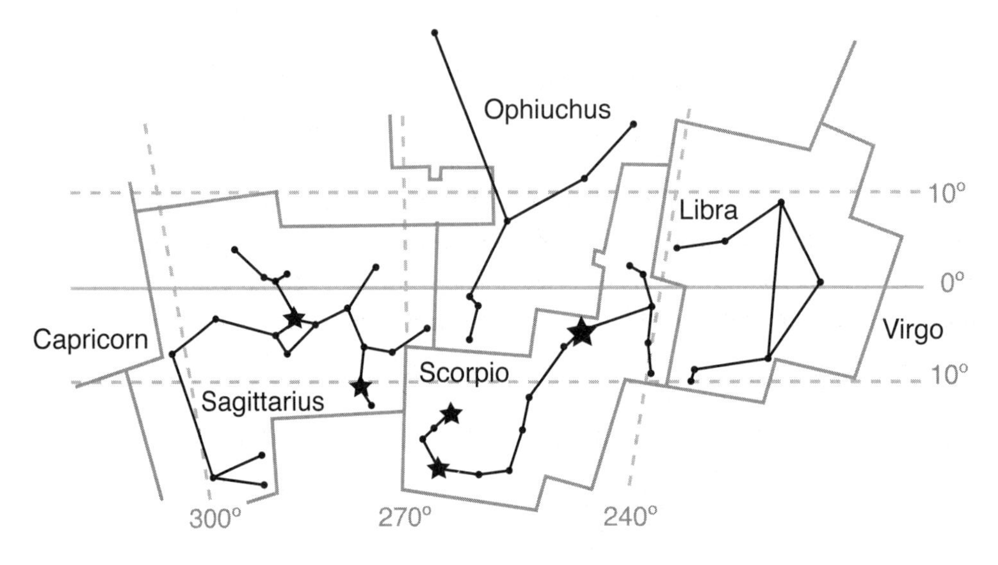

Figure 28. Libra, Scorpio, Ophiuchus, and Sagittarius (right to left), around midnight on June 20, or around 9 p.m. on August 1, low in the southern sky.

There are no very bright stars in Libra, so unless you're in a truly dark location, first find Scorpio's Antares and Virgo's Spica. Libra's brightest star, Zubenelgenubi ("the Southern Claw"), is between them on the ecliptic, 24° from Antares and 21° from Spica (see figure 29). As the name "southern claw" implies, this is Scorpio's claw star *within* Libra's scales; the fainter "northern claw," Zubeneschamali, is nearby. In some old illustrations, these claws hold the balance pans of the scales, an example of how fluidly our ancestors viewed their constellations—layered and overlapped rather than precisely regimented.

Wedged in on Scorpio's left is Ophiuchus, the Serpent Bearer (see figure 28). There's more to this tall figure than is shown here, but note that the Serpent Bearer occupies a good chunk of space along the ecliptic, about 19°. Ophiuchus is an example of a topic that puts astronomers and astrologers at odds. The former see the *fact* of thirteen constellations along the ecliptic, while most of the latter work with the traditional twelve signs.

To the left and below Ophiuchus is Sagittarius, the Archer (see figure 28). Bright stars Nunki and Kaus Australis help to form the Teapot, an asterism within Sagittarius that is easier to locate than the constellation as a whole. From

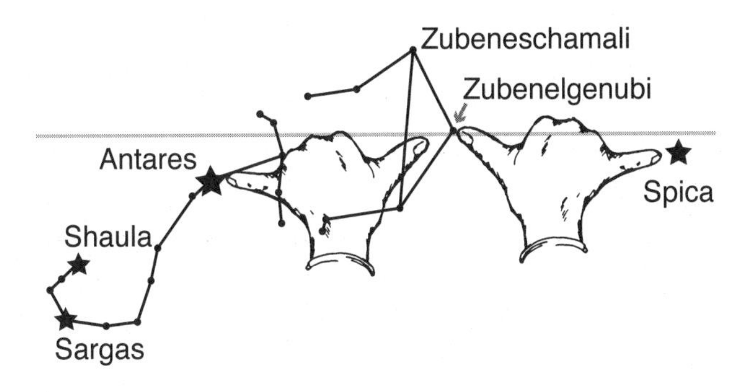

Figure 29. Finding Libra.

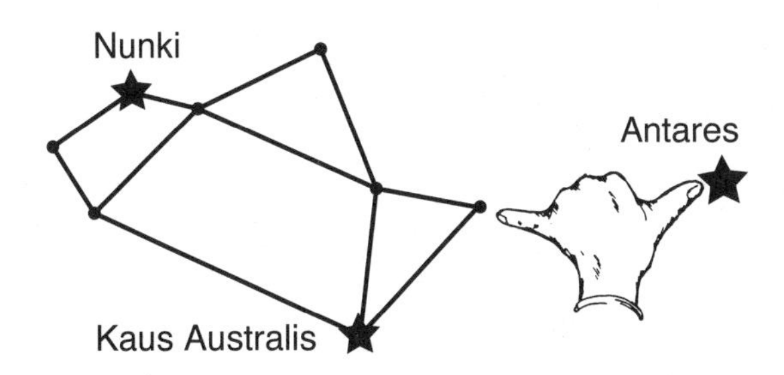

Figure 30. Sagittarius' Teapot asterism.

Antares, look left to find the tip of the Teapot's spout 21° away (see figure 30). From a dark enough location, you'll also see the Milky Way running up through Scorpio, Ophiuchus, and Sagittarius.

The Autumn Equinox showcases—quietly—the final three signs of the zodiac, Capricorn, Aquarius, and Pisces (see figure 31). These contain no especially prominent stars, so look high up for the Square of Pegasus, an asterism within the large constellation of Pegasus, and work outward from there (see figure 32). Using the Square's Alpheratz-to-Markab diagonal, head directly out a little more than 20° to Aquarius the Water-Bearer's right-shoulder star, Sadal Melik. Continue along that line to Capricorn.

Returning to the Square of Pegasus, note the angle between Alpheratz and Algenib. Now head to the left across the sky and find Hamal, in Aries. We've come full circle. Using your hand span, find Alrescha, the tip star in Pisces, 20° from Hamal at the same angle as that between Alpheratz and Algenib (see figure 33).

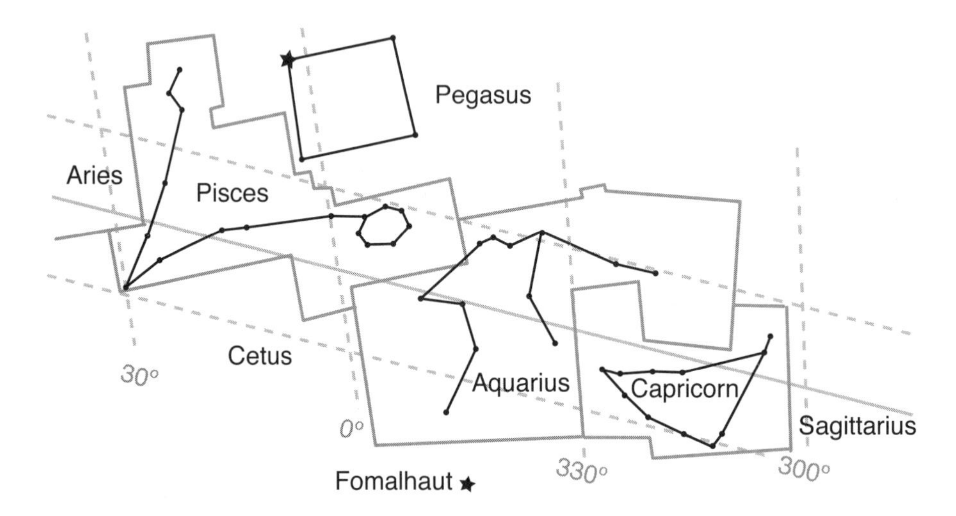

Figure 31. Capricorn, Aquarius, and Pisces (right to left), visible around midnight on September 20, or around 9 p.m. on October 31.

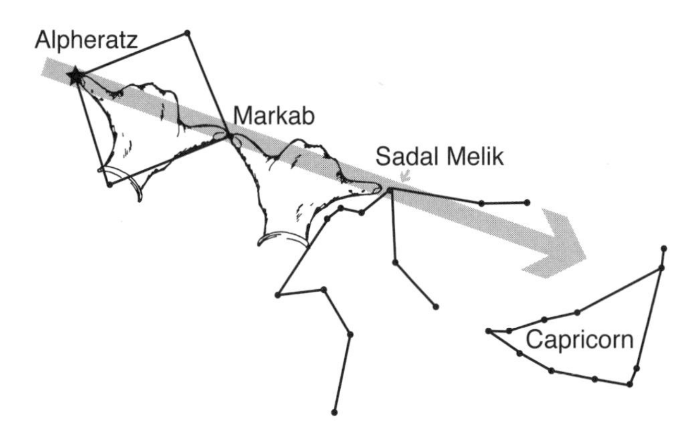

Figure 32. Finding Aquarius and Capricorn using the Square of Pegasus.

Alrescha comes from the Arabian *Al Risha,* "the Cord," for the starry twine that threads the two Pisces fish together.

Look at figure 34 for some specifics on the zodiac constellations. Note first that the "Sun occupies" dates given on this chart are for the Sun's symbolic astrological position and then the Sun's actual position in the sky. I'm using the modern boundary lines, which *aren't* what cause the discrepancy between astronomical and astrological locations and dates, as we'll see in chapter 3.

Second, remember that the meridian is an imaginary line running overhead from the northern horizon to the southern horizon. Use a compass to check this, or simply find Polaris and then stand with it at your back, facing south. Two signs can occupy the midnight meridian simultaneously. Refer to the preceding pictures to see how they can overlap.

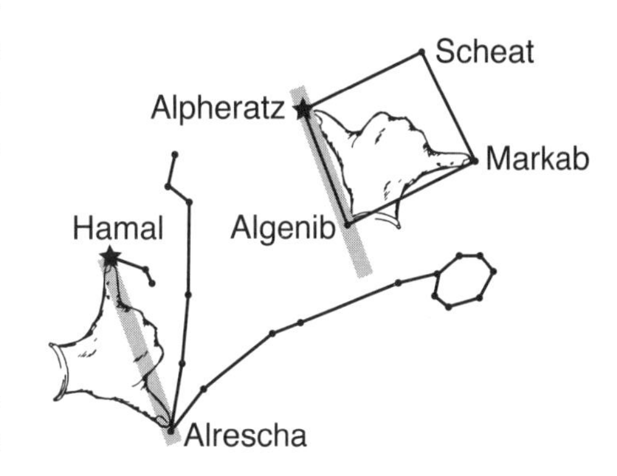

Figure 33. Finding Pisces using the Square of Pegasus.

Aries, the Ram

The zodiac traditionally begins with Aries, in honor of its former role as the location of the Spring Equinox Sun from c. 1850 BCE to c. 70 BCE (more on this in chapter 3).

At around the same time, the Egyptian God Amun came to prominence (Middle Kingdom, 2000–1786 BCE). Amun was creative and fertile, but He was called "the hidden one" because He was invisible. The atmosphere god Shu urged Amun to show Himself, so Amun donned a ram's skin and head.[26] He was thus "made visible" as a small constellation that gradually became the equinox point.

Among Amun's gifts to humanity were the orderly motion of time and the seasons, the better to nourish all the creatures and plants that His fertility created. One of His titles was "protector of the road," which befits Aries' role as the Spring Equinox constellation. There are rows of ram-headed sphinxes at Karnak and many other ram expressions in Egyptian art. As another example of life spun into existence, we have the ram-headed deity Khnum, who created children and their *ka*—the individual soul of each child—from celestial clay on His whirling potter's wheel.

An early Greek tale tells of two children, Phrixus and Helle, whose mother was a cloud goddess named Nephele. When the children were threatened, Nephele sent a magical, winged, golden-fleeced ram to rescue them. The ram's

Symbol	Zodiac Sign	"Sun occupies..." and midnight meridian dates (MM)
♈	Aries, the Ram	Astrological: March 21 to April 20 Actual: April 18 to May 13 MM: all through November
♉	Taurus, the Bull	Astrological: April 21 to May 20 Actual: May 13 to June 21 MM: late November to early January
♊	Gemini, the Twins	Astrological: May 21 to June 20 Actual: June 21 to July 20 MM: early January to early February
♋	Cancer, the Crab	Astrological: June 21 to July 22 Actual: July 20 to August 10 MM: early to late February
♌	Leo, the Lion	Astrological: July 23 to August 22 Actual: August 10 to September 16 MM: late February to early April
♍	Virgo, the Maiden	Astrological: August 23 to September 22 Actual: September 16 to October 31 MM: end of March to late May
♎	Libra, the Balance	Astrological: September 23 to October 22 Actual: October 31 to November 22 MM: mid-May to early June
♏	Scorpio, the Scorpion	Astrological: October 23 to November 22 Actual: November 22 to November 29 MM: early June to early July
⛎ or ⚕	Ophiuchus, the Serpent-Bearer	Astrological: not included Actual: November 29 to December 17 MM: early June to mid-July
♐	Sagittarius, the Archer	Astrological: November 23 to December 21 Actual: December 17 to January 19 MM: beginning of July to mid-August
♑	Capricorn, the Goat-Fish	Astrological: December 22 to January 19 Actual: January 19 to February 15 MM: early August to early September
♒	Aquarius, the Water-Bearer	Astrological: January 19 to February 19 Actual: February 15 to March 11 MM: mid-August to early October
♓	Pisces, the Fishes	Astrological: February 20 to March 20 Actual: March 11 to April 18 MM: mid-September to early November

Figure 34. Zodiac chart showing the Sun's position relative to the zodiac signs.

golden fleece lives on in the story of Jason and the Argonauts, but its rescue of the abused children is the more poignant image that lingers.

Taurus, the Bull

This constellation—Mul Gu-An-na, or "the Bull of Heaven"—represented abundance and prosperity, like the animal itself. Cattle were valued for their meat and hides; but around 4000 BCE, their uses expanded. They came to be valued for muscle power and dairy production, which meant that the animals were retained and cared for rather than simply hunted and eaten. Even now, in some parts of the world, stable subsistence living means owning two head of large cattle. With them, you can plow and have milk, and their dried droppings provide enough cooking and heating fuel for a small dwelling year round.

The Egyptians believed that the human race was created when the Sun was in Taurus at the Spring Equinox—c. 4540 BCE to c. 1850 BCE.[27] The goddess Bat, one of their earliest deities, was portrayed with a cow's face. Bat evolved into Hathor, and this goddess is depicted with a cow's head or as a woman wearing a cow-horned Sun headdress. Her worship predates the written Egyptian language, meaning that Hathor was present before c. 3400 BCE. Hathor's realms were childbirth and motherhood, music and dance, joy and love. These aren't associations most of us have with cattle, but the loss is ours. Family dairy farmers in particular have a symbiotic relationship with their herds; they tend to know their animals by name and understand that the volume and quality of milk depends upon the animals' contentment. Judging by the spray of Her milk, which created the Milky Way, Hathor must have been *very* contented.

Gemini, the Twins

Castor and Pollux are stars of similar magnitude—a steady side-by-side team and a good inspiration for the ancient themes of both "twin" and "best friend."

In the Sumerian tales of Gilgamesh (c. 2150–2000 BCE), wild-man Enkidu is Gilgamesh's "second self." Gilgamesh was the mythic semi-divine king of Uruk, a being who believed himself without rival in skill and strength. Hubris set in, however. He thought himself infallible and began to oppress his people brutally.

The gods countered this by creating a rival, the wild-man named Enkidu, who was Gilgamesh's opposite in many ways—uncultivated in manners and learning, rough and shaggy in appearance. When Enkidu was first found running wild, a "temple harlot"—or *hierodule*, a temple priestess of sacred sexuality-—was brought to the forests to encounter him and hopefully to transform him through human intimacy. The woman was Shamhat, and her impact on Enkidu was powerful. He submitted to her, and their sexual encounter tamed his wildness while transmitting knowledge and understanding. Shamhat then escorted Enkidu into the city, where he could meet the only man worthy of his companionship: Gilgamesh.

Gilgamesh, meanwhile, had prophetically dreamed that he would soon meet a man of great strength who would become his dearest ally and friend.

When Enkidu arrived in the city, he initially fought Gilgamesh. As king, Gilgamesh claimed the right to ravish each new bride on her wedding day before she could join with her husband. Enkidu arrived during a wedding and was appalled to learn of the practice. In the bride's defense, he challenged Gilgamesh and the two fought, soon proving themselves evenly matched in skill and strength, confirming that each was the most worthy companion for the other. More adventures followed, of course, but an ironic component of this story is the contrast between the men. Gilgamesh, who seemingly had everything else, lacked wisdom and compassion and the ability to govern his own impulses; Enkidu, the rough-hewn "wild-man," was transformed by intimacy with one woman and then came to the defense of another. Enkidu brought a civilizing influence to Gilgamesh.

The theme of twinship was common in the ancient myths. In the Babylonian pantheon, for example, Inanna and Utu aren't just sister and brother; they are also twins. Egyptian Earth-god Geb and sky-goddess Nut are both twins and lovers; among their children, twin sons Osiris and Set are bitter rivals who are married to their twin sisters Isis and Nephthys, respectively.

Many other cultures have important Sacred Twin myths. These include Romulus and Remus (Roman); hero-twins Hunahpa and Xbalanque (Maya);

solar male Liza and lunar female Mawu (African, Fon tribe); Moon goddess Artemis and Sun god Apollo (Greek); Monster-Slayer and Child-of-Water (Navajo); Shachar, "dawn," and Shalim, "dusk" (Canaanite). There are many other mythological twins, but these are some of the earliest.

The twins Castor and Pollux, for whom Gemini's two most prominent stars are named, are semi-divine. Taking the form of a swan, Zeus came to the mortal woman Leda and seduced her; Castor and Pollux arrived in an egg. Although these two stars are always treated as a pair, they aren't always perceived as human. The Arabs saw Castor and Pollux as a pair of peacocks; the Maya saw them as two peccaries copulating.

Cancer, the Crab

This crab makes a modest appearance in Greek myth. While Hercules wrestled with the Hydra (preserved as the constellation just below Cancer), Hera sent the Crab to distract him. Hercules was the son of one of Zeus' mortal consorts, and Hera took every opportunity to challenge the semi-divine hero. Crab's role was brief—Hercules crushed it underfoot—but Hera placed the small crustacean in the heavens in appreciation for its efforts.

Cancer's importance comes from its location—it's on the ecliptic—and its identification with summer. The Sun was in Cancer at the time of the Summer Solstice from c. 1500 BCE to c. 1 CE. The Akkadian word for "crab" was *alluttu*, but a crab could also be called *Kushu*—water creature—a more inclusive word that also encompassed turtles, which have the same rounded shape.

A pre-Hercules Mesopotamian fighter named Ninurta had many adventures that were later paralleled by Hercules. In his exploits, Ninurta battled a seven-headed serpent and then a turtle. The gods of his time didn't see Ninurta as a hero, however; they considered his motivations selfish. Ninurta defeated the serpent but couldn't overcome the turtle, which had been created by the god Enki specifically for this battle. Turtle and warrior fell into a pit, where they continue to struggle with each other.

Early Egyptians (c. 4000 BCE) saw Cancer as a tortoise, but later records call it a scarab.

Leo, the Lion

Many cultures see a feline shape in Leo, which is sometimes drawn with the lion crouching, as if ready to pounce. More often, however, Leo is shown in the standard pose of a contented cat surveying the world with paws extended comfortably.

Many goddesses are associated with lions and other large felines. One of the earliest interpretations is a sculpture (c. 7100–6300 BCE) found at Çatal Höyük in Turkey. Here, a woman is shown seated on a throne with a large feline on each side. She is heavily built, with a commanding presence. On close examination, the throne is seen to be a birthing chair, with the head of the woman's child just beginning to emerge below her. The felines are the height of armrests and in the same position, as if supporting her through the birthing process. The figure is crafted of terra cotta clay and was found in a granary (for abundance and fertile fields).[28]

Was the seated woman a goddess? The people left no written description, but her demeanor and regal pose speak volumes. The cats with her are often described as leopards, in part because other feline depictions from the same region are shown clearly marked with leopard spots. What leopards meant to the people of Catal Höyük we're unlikely to discover, but other cultures have draped their priestesses and priests in leopard skins, taking the animal's spots as representations of the starry night sky.

There are other goddesses accompanied by felines as well. Cybele, whose worship stretched through centuries and across Old Europe, was often depicted in a chariot pulled by lions, or seated on a throne with a lion on each side, exactly like the Catal Höyük terra cotta. One center of Cybele worship was Lyon, France, where the basilica dedicated to the Virgin Mary—with winged lions guarding the entrance—now occupies the site of Cybele's huge temple.[29]

Among the Egyptians, Sekhmet is the lion-headed goddess (see figure 49). She is powerful both as a ferocious destroyer and as an agent of healing, especially from fevers. As a summer constellation, Leo and its lion form logically relate to heat.

In India, Durga is the deity associated with the big cats, and this Mother Goddess is often shown riding either a lion or a tiger, calmly subduing the demons of suffering and misery.

Virgo, the Maiden

Since ancient times, this constellation has been portrayed as female and associated with plants, harvesting, plowing, and abundance. The Babylonians saw the western half of Virgo as Erua, wife of the high god Marduk and a principal goddess. She's shown carrying a frond of the date palm in a distinct constellation called the Frond of Erua. Dates ripened in early autumn, which is when Erua and her frond appeared in the eastern sky just before sunrise. For the Babylonians, the eastern, later-rising portion of Virgo was The Furrow, depicted as a goddess named Shala who carried a stalk of barley. Remember, at a hefty 44°, Virgo is the widest constellation of the zodiac, so splitting its wide-stretched stars into two separate figures isn't unreasonable, especially since barley bread and dates were both dietary staples worth honoring. Later peoples managed to bring Erua and Shala together into the huge grain goddess we know now.[30]

Grain Goddesses, Great Goddesses, Mother Goddesses—all these deities were recognized very early in history and widely venerated. And why shouldn't Virgo extend across a vast area of the sky? She symbolizes the Goddess's powers to feed the people.

Many of Virgo's goddess attributes carried over to the Christian Marys, whether we interpret that as the Virgin Mary or as Mary Magdalene. By all Her various names, Virgo is the Great Goddess.

Libra, the Balance

In some old traditions, this constellation was called Pluto's Chariot, and was seen as the chariot in which Pluto carried off or pursued Proserpina.[31] This makes good sense because the constellation's shape echoes that of a cart and its tongue, and, in the mythology of Pluto and Proserpina (or the Greek Hades and Persephone), winter comes when the maiden is abducted to the Lower World. The Sun's entry into Libra starts to signal the arrival of winter, when the Sun is indeed much lower in the sky, as if diving down into the Earth itself. At sunset, Virgo-maiden Proserpina vanishes down over the western horizon right after the Sun, with Libra—or Pluto in his chariot—chasing right behind her.

Since Libra's scales represent a piece of machinery, I expected them to be a recent constellation construct, but that's not the case. Libra came in early. Its

Akkadian name was Mul Zib-ba-an-na, for "stars: a set of weighing scales." A beam scale—basically a stick with a pan suspended on either end, like the one the goddess of Justice carries—is the simplest version and the one closest to early depictions of Libra. To the Babylonians, the weighing scale was sacred to the Sun god Shamash, who illuminated all things and therefore was associated with truth and justice.

Scorpio, the Scorpion

This constellation's distinctive, hooked tail assured its ongoing identification with a scorpion among Sumerians, Babylonians, Egyptians, and other early people, many of whom—as dwellers in or near deserts—routinely saw real scorpions. As sky lore was shared among people who had no firsthand experience with scorpions, descriptions of the creature apparently made a hazy but horrific impression and the name Scorpio was retained, even when their depictions looked more like maniacal lobsters.

For the Egyptians, Antares represented Selket, their scorpion-crowned goddess. Despite her terrifying headdress, Selket gave special protection from poisonous bites and stings, and was the patroness of childbirth and nursing. The Egyptians considered scorpions to be good mothers. This image may fail to resonate for us—it's not a likely Mother's Day card motif—but the Egyptians were careful observers of nature, and they were correct. Mama scorpion carries all her tiny scorpion babies around on her own body, tending to their needs. Temples that were dedicated to Selket around 3700–3500 BCE had equinox sunrise alignments with the red star Antares.[32]

Usually Selket is portrayed as a beautiful human woman with a large scorpion perched on her head. A Selket figure was discovered guarding a corner of King Tutankhamen's sarcophagus, with her arms extended protectively like a petite gold-leafed crossing guard.

Ophiuchus, the Serpent Bearer

Although Ophiuchus is included here because he occupies space along the ecliptic, he comes up blank in many standard zodiac categories—with no assigned plants, attributes, elements, or modes and no space in the daily astrology column. Ophiuchus is pronounced o-FEW-kus. A snake wrapped around a staff or a large U with a superimposed tilde are two proposed symbols (see figure 35).

This large character is identified with Asclepius, a son of Apollo, who learned the healing arts from his father and the centaur Chiron. Ophiuchus' giant serpent stretches away to either side of him as two separate constellations: Serpens Caput ("the Serpent's Head") to the west (right); Serpens Cauda ("the Serpent's Tail") to the east (left).

Ophiuchus isn't wrestling this snake; it's his animal ally. A snake helped to teach Asclepius his vast skill in medicine, instructing him in the healing powers of the plant world. As a result, Asclepius' own symbol is a cane with a single snake spiraling up it. This came to be known as the rod of Asclepius, although it is often confused with the caduceus, as we'll see in chapter 5.

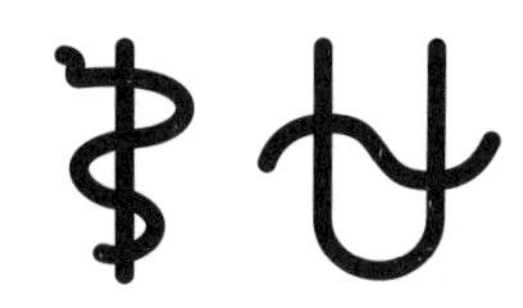

Figure 35: Two possible Ophiuchus glyphs.

A close association between humans and snakes often indicates magical or other special powers, and Asclepius was no exception. His healing skills became so powerful that he was able to cure every human wound and ailment, and could even reverse death itself. This accomplishment brought him to the attention of Hades, ruler of the underworld and receiver of the dead. When Hades' "clients" (the dead) suddenly stopped arriving due to Asclepius' intervention, the Lord of the Underworld complained to His brother Zeus, who struck Asclepius with a thunderbolt and then placed the healer and his snake ally together in the heavens. Lesson: Ultimate power over life and death is reserved for the gods. Rather than continuing to call Asclepius by his proper name once he was relegated to the heavens, he is now known as Ophiuchus, the Serpent Bearer.

Look at the sky instead of the daily horoscope and you'll find Ophiuchus and his snake right there on the ecliptic among the zodiac constellations. There's no point in ignoring him, but astrology—which is an art, not a science—functions well with the twelve signs it already has. However, given the beleaguered state of Earth and her occupants, isn't it significant that this mythic healer's constellation has now entered our awareness so dramatically? An emerging consciousness of healer Ophiuchus can be applied to the Earth challenges of our time.

Sagittarius, the Archer

To the Babylonians, Sagittarius was Pabilsag, who was often shown with great wings, a horse's legs, and a scorpion's tail. Pabilsag's scorpion tail may have come from the constellation we now call Microscopium, which is located within the back legs of the Sagittarian centaur. Pabilsag sometimes has a small dog on His back or at the back of His head, as if watching the road behind Him. Some of the earliest lore connects Pabilsag with Ninurta, who eternally wrestles with the Cancer turtle. Other stories unite Pabilsag with the healing goddess Gula as wife and credit Him with the role of psychopomp, a guide for the souls of those departing from life.

Pabilsag, or Sagittarius, is located just along the Milky Way. His bow is drawn, and His arrow points into the stream of stars. Sagittarius rides low on the horizon, so the Milky Way seems to begin right under the centaur's front hooves before arcing up into the sky. Gavin White, author of *Babylonian Star-Lore*, equates the hunter Pabilsag with the European theme of the Wild Hunt, a hard-galloping troupe of unquiet souls who travel the night winds between the realms of Earth and spirit—the living and the deceased.[33] The name *Pabilsag* means "Chief Ancestor," who might be expected to know the way into the next world and to lead the procession of souls going there.

The zodiac circle of animals contains few humans—in fact, just the Gemini twins, the maiden Virgo, and Aquarius the Water Bearer. The Sagittarian centaur straddles this category, being part human and part horse. In Greek myth, Chiron was a son of the deity Cronus and a mortal woman named Philyra. Cronus came to her in the form of a horse, and Philyra birthed a horse-child nine months later. Chiron grew up to be king of the centaurs, and a masterful hunter and healer renowned for his compassion and wisdom. He was friendly toward humans (unusual for centaurs) and, as a teacher, was responsible for raising and training many human and semi-divine boys, including Achilles, Aeneas, Asclepius, Hercules, Jason, and Peleus, all of whom went on to roles in their own noteworthy tales.[34]

The hybrid horse-human identities of both Pabilsag and Chiron can be interpreted as shape-shifting—the step into shamanism and trance that occurs when a human journeys beyond physical form into the realm of spirit and then successfully returns with useful information. In this between-the-worlds place—

symbolically, where Sagittarius steps into the Milky Way—the journeyer meets helpers from other species, perhaps taking on some of their animal qualities and bodily form. Seen from this perspective, Chiron's skill as a healer derives from his spirit connections with the horse, part of a wider tradition that views the horse not just as a means of earthly travel, but as a means of accessing the realm of spirit as well. The horse's hoof beats are replaced—or mimicked—by drumbeats. The knights in the tarot deck are a remnant of this theme, as tokens of both physical travel and those inner journeys by which we grow in comprehension.

Wounded healer is a phrase some modern writers have applied to Chiron, and this bears explanation. Chiron was accidentally wounded by a shot from Hercules' bow.

In shamanic terms, a "wounded healer" is someone who has been wounded—by illness, injury, or life's brutal circumstances—and has found healing through allies on the spirit plane. In this context, healing is distinct from curing. *Curing* refers to a physical shift—mending. *Healing* refers to spiritual balance and wholeness. For the "wounded healer," the healing may be spiritual, emotional, psychological, physical, or some combination of these; most crucially, however, healing has occurred and it occurred *through spirit*. This momentous occurrence and all its practical meaning and metaphoric weight are sometimes abbreviated to the mistaken idea that the wound itself is the main event—i.e. I am wounded; therefore, I am a healer. Not so. If we're showing off our hideous wounds thinking they convey power, we've missed the point. Healing through spirit is what matters.

Chiron had a profound understanding of healing, but even he couldn't heal himself from the poisoned arrow shot by semi-divine Hercules. Being semi-divine and immortal himself, Chiron also couldn't die until Zeus granted him mortality—literally, the ability to die. When Chiron left the earthly plane, Zeus placed him among the stars as Sagittarius.

Capricorn, the Goat-Fish

Aries is specifically a ram, a male sheep; Capricorn is half goat and half fish, which is different. We just saw the human-horse merger that is Sagittarius, and now we have a fusion of mammal and aquatic vertebrate.

Capricorn has been represented as the Goat-fish for at least four thousand years. This strange creature—the front half of a goat joined with the tail of a fish—appears in early Babylonian illustrations and continues in identical form with the Greeks. One school of thought connects the odd figure with the Babylonian deity Ea, who appeared out of the subterranean waters as an antelope to bring all the skills and knowledge of civilization to the people. In this scenario, the fishtail remains as a token of Ea's sojourn in the ocean.

In 2000 BCE, the Winter Solstice occurred just after Capricorn began to be visible in the southeastern predawn sky, the front (goat) half appearing first. This was—and is—a time of transition, as this half-and-half creature certainly expresses.

Although the Babylonian tales came far earlier, Greek lore evolved a new explanation for Capricorn. Their goat god Pan—human from the waist up, but with goat hind legs and well known for lusty mischief and wild escapades—underwent a similar transformation. When they were attacked by the monster Typhon, all the Olympian gods fled to Egypt, looking for ways to hide. The Grecian gods shape-shifted: Zeus became a ram (Amun), Artemis a cat (Bast), Hermes an ibis (Thoth), and Hera a white cow (Hathor). Pan jumped into the Nile. He kept His upper body above water and it remained unchanged, but His goat legs were submerged and became a fishtail.

Aquarius, the Water Bearer

To the Babylonians, Aquarius was Gu-la, the Great One, a towering human figure with a water jug in each hand, pouring out streams of water onto the Earth. In the old depictions, the Great One was taller than mountains; sometimes the water flows into a stream, and sometimes into a different pair of jugs.

Complicating this identity is another Gula or, more likely, the same being in another form. This other Gula was a goddess of healing and wife of Pabilsag, and some depictions of the Aquarian water-pouring figure are clearly female. In 1990, an archaeological dig in Nippur, Iraq, revealed a structure roughly the size of a football field that was used as both a temple and a healing center dedicated to the goddess Gula. The layers at the dig ranged in date from c. 1600–1200 BCE back to 3000 BCE.[35]

There were two kinds of healers in Babylon: the *asu*, an herbalist who also mended wounds and performed surgeries, and the *ashipu*, who worked from a spiritual perspective, diagnosing problems and then doing the appropriate rituals to assure a full recovery. The *asu* and *ashipu* worked together rather than competing. Gula's companion animal was a dog—thousands of small dog figurines were found at Nippur—which is intriguing given recent medical trials using dogs trained to sniff out cancer and other diseases. At least one temple inscription refers to the healing goddess Gula as the Great One, our constellation Aquarius.

Pisces, the Fishes

Where we see two fish separated in the sky but connected by a long cord, the Babylonians saw a bird and a fish. They were still connected by a cord, but the two creatures were viewed as two separate constellations. On the western edge of our Pisces was the Swallow, whose written name can also be translated as "Exalted Bird."

Pisces' eastern fish, which sits higher in the sky above the ecliptic, was written in Akkadian as Anunitum, the Star of the Goddess of Heaven, and is interpreted as being the Akkadian version of Inanna. The city of Akkad (on the Tigris, possibly near modern Baghdad) was the center of Anunitum's worship. Some images portray Anunitum as a mermaid. Given her lofty title as Star of the Goddess of Heaven, White theorizes that this constellation once included Andromeda as the Goddess of Heaven just above Pisces' eastern-most fish.[36]

Pisces is directly opposite wide Virgo in the sky—a great goddess related to water and fishes on one side of the heavens; a great goddess of the grains on the other side. This has a pleasing symmetry.

Where Astrology Comes In

I studied astrology as a teenager, but only years later learned how to find my own Sun-sign constellation in the night sky. This is sadly common. Astrology in our times has little to do with actual observation of the heavens. Back when my grandmother taught me the basics of "casting a chart"—creating a sky map of the zodiac and planets for a particular time and place—we used an ephemeris, paper, and pencil. Period. (That's when I realized that I liked math, at least for this application.)

Eventually, calculators helped speed up this procedure, and now computer programs make astrological chart-casting painless—although more detached than ever. We aren't going to look much at charts here, since that's astrology and no longer an exploration of the sky. However, there are some key concepts found in astrology's symbolic use of the zodiac that are intimately tied to many other metaphysical systems. For ease of learning or remembering them, use the chart of zodiacal constellations (sorry, Ophiuchus) and their glyphs that we saw in figure 34. Learn to recognize the glyphs! They show up plentifully elsewhere—worked subtly into tarot card designs, for example—and if you can't read them, you'll miss layers of meaning.

Early associations of each planet with specific zodiac signs can be traced back at least to al-Biruni, the brilliant Persian Muslim scholar (973–1048 CE). The signs were parceled out among the planets in a very orderly manner: Leo went with the Sun, Cancer with the Moon, and the rest simply fell into place in keeping with their order in the sky (see figure 36).[37]

As the zodiac travels into other metaphysical systems, the most pervasive and consistent concept is that of the elements—the idea that each of the twelve signs resonates to earth, air, fire, or water. We've all heard people say, "I'm an Earth sign" or "She's a Fire sign." This is the origin of their remarks. Each element gets three zodiac signs, so these groups are sometimes called "elemental triads" (see figure 37). Like so many things in metaphysical studies (and life in general), if this element business doesn't make logical sense to you right off the bat, give it time to sink in. The elemental associations tend to grow in your awareness and can become profoundly meaningful.

If you have an interest in tarot, you'll find the elements waiting for you there in the four suits of the Minor Arcana, although the assignments vary between decks. I use a system in which Swords are associated with Air, Wands with Fire, Cups with Water, and Pentacles with Earth. Some decks associate Swords with Fire and Wands with Air, while other decks scramble the element/suit pairings in different ways. Which one is correct? Whichever makes the most sense to you personally. Both tarot and astrology benefit from a core of empirical learning, but ultimately—in the actual application—each is rooted in the heart-connected intuition of the practitioner.

Modes are another way of viewing the zodiacal signs in groups—in this case, three groups of four signs each, one from each element (see figure 38). Here's a good way to remember these combinations in astrological terms:

The Cardinal signs are those that begin with a solstice or an equinox.

The Fixed signs are those sitting in relative stability between Cardinal and Mutable signs.

The Mutable signs are those about to change, as they immediately preceed a solstice or an equinox.

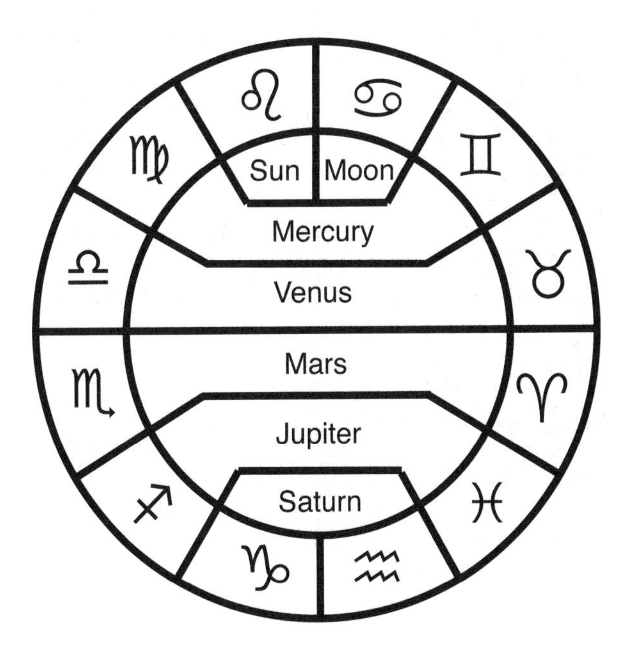

Figure 36. The al-Biruni circle of the zodiac and the planets.

Figure 37. The zodiac signs and their associated elements shown as elemental triads.

Figure 38. The zodiac signs and their associated elements shown as modes.

Learn by Doing

1. **Find your Sun sign:** Learn to locate your own Sun sign—where the Sun was when you were born. Check the chart in figure 34 for midnight meridian periods (MM), when visibility should be good and the Sun far from the stars you're trying to see. This is a means of personalizing the night sky. Children find this fascinating, with or without an accompaniment of old lore. Moonless nights offer the greatest visibility.
2. **Orient yourself to the heavens:** The next time you're noticing stars, check with the sky charts at the beginning of the chapter to see if some zodiacal stars are involved. Like memorizing landmarks in a new town, this is a way to get your bearings among the heavens.
3. **Explore the tarot:** There are many Tarot systems for equating each sign of the zodiac with a Major Arcana card. Some are better known than others, but if you want these correlations, explore your options and see what feels like the best fit, or devise your own correspondences.
4. **Become a part of the dance:** The ecliptic, the Sun's path through the sky, follows the stars we've been working with here. As you begin recognizing these stars in the sky, notice how the ecliptic tilts—gauged by how high or low these stars are above the southern horizon—moving from day to night, and from season to season. The gentle to-and-fro swing of this tilt is just part of the stately and elegant dance of the heavens. As you come to recognize the rhythm, you can remember on a deeper level that you're part of this great dance.

Chapter 2

The Dance of the Sun

We'll be exploring Sun lore next, but first, let's get our bearings. So far we've been looking at the stars, poetically called the "firmament" because, although they rise and set and move overhead, they look reasonably fixed (or "firm") in their positions relative to each other. The ancients saw roughly the same star arrangements, whether or not they connected the dots in quite the same way.

Planets are different from stars. No firmament here. The name itself comes from the Greek *planetai,* which means "wanderers." In contrast to the stars, the planets buzz around like go-carts, all at different speeds; some even seem to go backward at times. We gauge their motion by the shifting backdrop of the zodiacal constellations behind them.

For our purposes, the "planets" are the seven "wanderers" visible to the naked eye—the Sun, the Moon, Mercury, Venus, Mars, Jupiter, and Saturn. We're skipping Uranus, Neptune, and Pluto because they aren't visible to the naked eye. The Sun and Moon aren't actually planets—the former is a star and the latter is Earth's satellite—but for ease of discussion, they're all grouped together here.

Besides their unique status as star and satellite, the Sun and Moon are unlike the other wanderers in other ways as well. As expressed through myths and markers, there's an inherent understanding that these two celestial bodies have a unique relationship, to each other and to the Earth. First, they each light up our world—one by day, the other by night. Second, they're nearly identical in size from an earthly perspective, and look huge compared to the other planets. Third, neither moves in reverse—retrograde—ever! Fourth, the Moon's shape-shifting nature is based on its position relative to the Sun. And fifth, the Sun and Moon work together to create eclipses.

Next on our list are Mercury and Venus. These two are "inner" planets—those that are inside Earth's orbit, in between Earth and the Sun—which affords special qualities to the motion of each.

Finally, in the "outer" planets—Mars, Jupiter, and Saturn—we see longer, slower cycles and other unique features, since these three planets are farther from the Sun than the Earth is.

How these planets move—the intricacies of their cycles—has historically been the stuff of record-making, calculation, and observation. The stories behind their motions, however, are the stuff of creative imagination, spirit, and inspiration. Both ways of knowing are deeply rooted in our human relationship with the sky.

The Sun and the Seasons

Night passes, and then the stars fade and the Sun rises. It travels across the sky and sets. This sequence repeats, over and over. Always the same.

Except that our view of the Sun changes daily.

We readily notice that the Moon changes its shape and appears at different hours of day or night through the course of its cycle. By comparison, the Sun's motion seems simpler and more constant: it comes up, goes overhead, goes down, repeat, repeat. But when we note that the Sun gradually changes its height in the sky and its position along the horizon, we mark the seasons—one of our most ancient acts of counting and timekeeping.

Many of us celebrate holidays by marking the solstices and equinoxes and the rough midpoints between them—eight holidays reflecting the seasons, like eight spokes in the turning wheel of the year. With central heating, electric lights, and supermarkets, we pay far less attention to the Sun than our ancestors did, but we notice the seasonal change as we reset our numerous clocks, forward or back.

Our early ancestors were clock-watchers too, but they used the Sun itself to chart the passage of time. And rather than adjusting it, they adjusted their own activities. They watched the Sun as if their lives depended on it—as, in fact, they did. If we're clock-watchers or basking Sun-worshippers now, we came by it honestly. For most of human history, time and sunshine have been life-or-death issues.

Scraping snow off my windshield on a winter morning, I'm cold, cranky, and forgetful of how easy I have it. Many of our ancestors struggled through at least one season of each year, whether it was a freezing winter, torrential rain, or brutal drought. They hoped that they'd stored enough food and fuel, they hoped their hunts would be successful, and they hoped . . . well, they just hoped, and worked, and traveled, and planned, and strived, and struggled. If the harvest was poor and the hunting meager, the goddesses and gods of the Afterlife would come looking for new recruits.

In some cultures, if rations grew short, a weakened elder voluntarily made a quiet exit into the night. In other cultures, grandma's exit probably wasn't voluntary. Some peoples optimistically called the place where souls went in the Afterlife "The Summerland," which speaks volumes about their winters of discomfort, potential deprivation, and dread.

Watching and understanding the sky, our ancestors knew when to stock their primitive larders, when to move camp to more congenial altitudes, and when to prepare to follow the herds. Survival of the fittest. Failure to understand the information provided by nature, coupled with plain old bad luck, removed many from the gene pool. We're the offspring of all the others—smarter, luckier, or simply more adaptable.

Yet sky watchers all.

The Sun's Motion

At the Summer Solstice in late June, the Sun rises in the northeast, soars high overhead at noon, then sets in the northwest. It holds to this extreme position for three or four days. In this season the days are long, the evening twilight lingers, and the nights are short, as are shadows. The noonday shadows on the

Summer Solstice itself are the shortest possible in the Northern Hemisphere, since the Sun is most directly overhead at that time.

But from the Summer Solstice onward, each morning at dawn as we face the east, the Sun rises a little farther to the right—the south—along the eastern horizon. Three months later, around September 20th at the Autumn Equinox, the Sun rises due east. *Equinox* means "equal nights," and sure enough, at the equinoxes, day and night are near-equal in length. Equinoxes are thus seasonal balance points (see figure 39).

From the Autumn Equinox onward, the Sun's shift becomes really obvious—ominously so for those of us who live in the northern latitudes. The Sun goes farther and farther south, and rides lower in the sky. The days get shorter, noon shadows are elongated and surreal, and the nights get longer and generally colder as well. Plants wither; animals migrate or hibernate.

Knowing just how far south the Sun would go was crucial. Besides the obvious practical considerations, the psychological impact was immense—it still is. Around the world, hundreds of sites incorporate Sun-watching markers. Many operate on a simple principle: if you stand here and look over there, you'll see the Sun rise or set over a special hill or between particular rocks, or illuminate an otherwise dark place.

As marked by whatever signpost each culture employed, when the Sun reaches that point, it stops moving south. What a relief! The Sun briefly holds its position, rising or setting in the same spot for three or four days. The Latin roots of the word *solstice* mean exactly that: "Sun stands still." Remember, the Sun is always shifting from north to south, south to north, *except* at the solstices, when it maintains the same position along the horizon for several days.

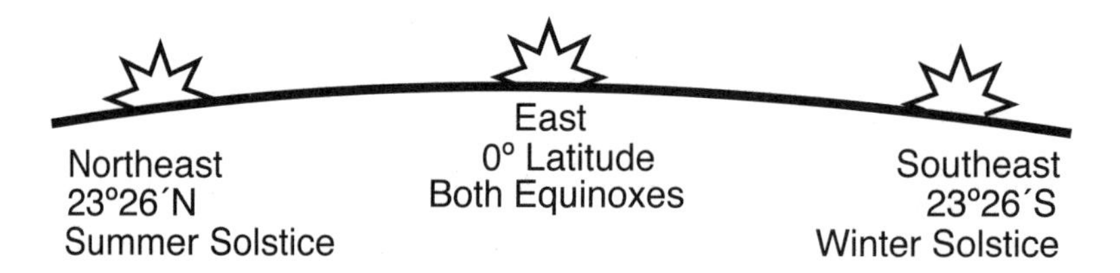

Figure 39. The Sun's shifting place along the eastern horizon with the equinoxes as balance points.

The Sun's southern-most declination—its north-to-south place in relation to the equator—is reached around the 19th or 20th of December.

We think of winter as a time when all is dormant. Not quite true. Trees carry out their root growth during the fall and winter months—a counterbalance to spring and summer, when they're busy with visible growth. Some animals gestate, following their fall mating seasons. Although travel conditions may be brutal, many animals—and humans—have relied on winter ice to traverse otherwise impassable waterways.

Then the Sun begins its gradual journey back toward the north, and the days slowly begin to lengthen.

Those are the mechanics of what occurs. From a spiritual perspective, the Sun Goddess or God returns north. Plants and animals awaken. We see the return of migratory birds and beasts. Slowly the Earth becomes lush and fertile. The days continue to lengthen and, finally, around March 20th, the Sun rises due east again. This is the Spring (or Vernal) Equinox, when days and nights are near-equal in length (the exact balance point actually happens a couple of days earlier).

From the Spring Equinox onward, days are longer than nights, until finally, on about June 19th or 20th, the Sun reaches its northern-most point and holds that position, "standing still" again for several days as we reach the Summer Solstice, back where we began. Looking for the dawn Sun on the eastern horizon, you must gaze well to your left—northeast rather than simply east. This seasonal extreme is also celebrated and aligned at sacred sites worldwide, with markers oriented to the Summer Solstice sunrise or sunset points—or both—on the local horizon.

The solstices and equinoxes are the four key moments in the solar year for the Earth, with the solstices being the most dramatic—thanks in part to our deep ancestral memories of perilous winters past. In simplified terms, we use the Summer Solstice to celebrate early harvest and to brace for what will come; we use the Winter Solstice to celebrate the return of the light—and we use the equinox point, the middle position shared by Spring and Autumn—to mark and honor the balance point. Many ancient cultures marked at least one of these points; some marked all.

At the Winter Solstice, that dramatic and dangerous time, the priests and priestesses, the medicine people, the shamans, the elders and wise ones, the rulers, and the people all watched the Sun. To see it reach its southernmost point, rise there for several days, and then begin edging back northward was cause for celebration. We still celebrate around that time, whether we call it Yule, Hanukkuh, Kwaanza, or Christmas.

An almanac or ephemeris gives the Sun's exact daily positions as calculated for a specific and consistent time and place, often noon at Greenwich, England, which is at 0° longitude.[38] The Sun's north-south location in the sky relative to Earth's equator is called its declination. To find the solstice dates, look for the Sun's standstill, the dates during which there's no shift in the Sun's northern- or southernmost position and at the highest-numbered declinations during the year. To find the equinoxes, look for the dates when the Sun's location shifts from north to south, or south to north. These numbers will be 0° or close to it. Figure 40 shows part of an ephemeris. Note the dates, the declination of the Sun, and the solstice and equinox points.

The Analemma

From our earthly perspective, gauged only by what we see along the horizon, we get an impression of the Sun moving forward and back, north to south, in straight lines—like an inexperienced driver struggling unsuccessfully to free a car from deep snow.

But the Earth moves through space in an ellipse, on a tilt. Without that tilt and ellipse, we'd move like that novice driver, in a straight-line rut. Nature, however, is inherently fluid and curved; so, instead of a rut, we get seasons and other elegant mysteries, including a graceful loop called the *analemma.*

The analemma is one way of observing the Sun's motion throughout the year. It's shown on some globes and maps, always positioned between the Tropics of Cancer and Capricorn, an expression of the Sun's journey between its maximum-north summer declination and its maximum-south winter declination (see figure 41). The northernmost point finds the Sun at 23.45° (or 23° 26') north latitude (directly over the Tropic of Cancer); the southernmost point is directly over the Tropic of Capricorn at 23.45° (or 23° 26') south latitude. The word

Month and Date	Declination of Sun ° ′	Motion and Season	
June 18	23 N 25		
June 19	23 N 26	Sun's	SUMMER
June 20	23 N 26	northern	SOLSTICE
June 21	23 N 26	standstill	
June 22	23 N 26		
June 23	23 N 25	← Sun begins moving south	
September 20	00 N 53		
September 21	00 N 31	Southbound	AUTUMN
September 22	00 N 06	← Sun	EQUINOX
September 23	00 S 15	crosses	
September 24	00 S 46	equator	
September 25	01 S 04		
December 18	23 S 25		
December 19	23 S 26	Sun's	WINTER
December 20	23 S 26	southern	SOLSTICE
December 21	23 S 26	standstill	
December 22	23 S 26		
December 23	23 S 25	← Sun begins moving north	
March 17	01 S 09		
March 18	00 S 47	Northbound	SPRING
March 19	00 S 24	← Sun	EQUINOX
March 20	00 N 01	crosses	
March 21	00 N 22	equator	
March 22	00 N 46		

Figure 40. A partial ephemeris, showing dates, the declination of the Sun, and the solstice and equinox points.

tropic comes from the Middle English *tropik,* which means "solstice," from the Greek *tropè,* which means "turn." This is literally where the Sun "turns around," and these Tropics are named for the Sun's entry into Capricorn (winter) and Cancer (summer) when these latitudes are reached. (Let's say "alleged entry": more in the next chapter.)

Photographers capture a visual record of the analemma by snapping shots for an entire year, always from the same perspective at the same time of day and

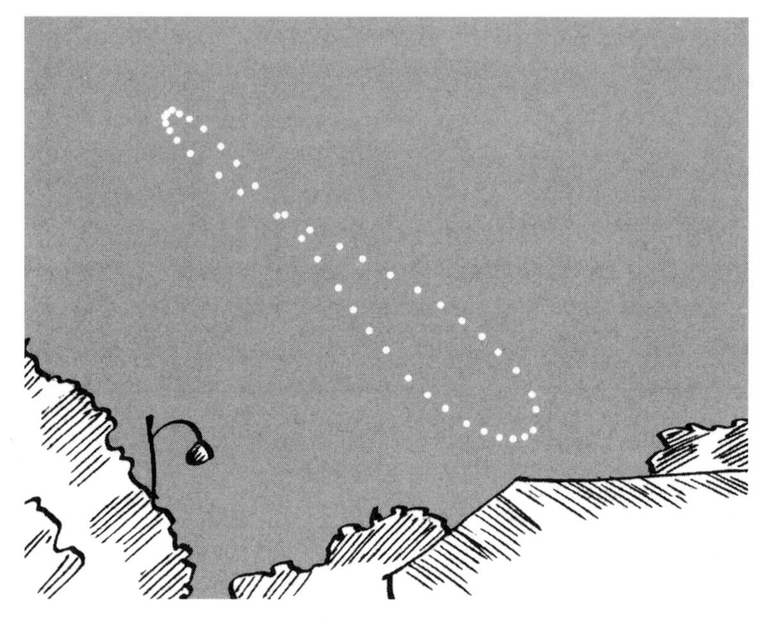

Figure 41. The analemma over a landscape.

from the same angle. The result shows a dotted line of multiple Suns shaping a giant figure-8 across the sky (see figure 41). The analemma's shape varies by location, but is seldom perfectly balanced.

The ends of the analemma represent summer and winter, but if we expect the solstices to be exactly at the ends and the equinoxes to be at the crossing point, we'd be wrong. Remember, tilt and ellipse. The Earth is tilted on its axis, and our orbit is oval—elliptical, rather than circular (see figure 42).

The analemma also manifests itself through shadows, as incorporated into many sundials. The shadow-casting portion of a sundial is an upright marker called a *gnomon.* Throughout the year, the tip of the gnomon's shadow traces the analemma's figure-8.

In 1655, English mathematician John Wallis used the analemma shape on its side to express mathematical infinity; in 1694, a Swiss mathematician named Jacob Bernoulli named this shape the "lemniscate" (see figure 43).[39] The shape is also used to express infinity in metaphysical contexts. The word *lemniscate* comes from *lemniskos,* Greek for "ribbon." A related Latin word, *lemniscus,* refers to a bundle of nerve fibers located in the brain. These fibers transmit discriminatory

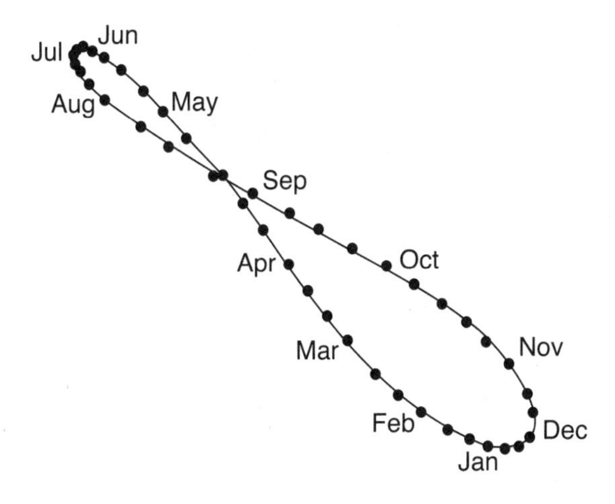

Figure 42. The analemma by date, describing a figure-8.

sensations from the skin to the cerebral cortex. In a sense, these are our own personal, sense-monitoring ribbons.

The Sun's analemma-lemniscate makes several appearances in the tarot. In the Minor Arcana, it is often depicted on the Two of Pentacles as a powerful image of balancing (see figure 44). In the Major Arcana, it traditionally appears on the Magician and Strength cards as a hat brim or hovering overhead (see figures 45 and 46). In the Rider-Waite-Smith deck, it is also shown as loops of ribbon on the wreath in the World card (see figure 47). These red ribbons and their placement are significant and appropriate, since another meaning for the Latin *lemniscus* is "ribbons attached to a victor's wreath."

Figure 43. The lemniscate as the symbol representing mathematical infinity.

Another expression of this shape is found in the Möbius strip, a narrow strip of paper formed into a loop, but rendered "endless" by incorporating a twist. The Möbius strip acts as if it has only one side—perpetual, like the lemniscate. In recent years, the Möbius strip has morphed into a shape that symbolizes recycling (see figure 48). We thus see it daily without recognizing it. Infinity in the most pragmatic sense.

I have a friend who calls the Afterlife "The Lazy-8 Ranch," a cosmic cowboy Heaven with a lemniscate cattle brand, of course.

Figure 44. The Two of Pentacles card from the Rider-Waite-Smith deck showing the analemma-lemniscate.

Figure 45. The Magician card with the lemniscate overhead.

Figure 46. The Strength card with the leminiscate overhead.

Figure 47. The World card in the Rider-Waite-Smith deck showing the lemniscate as loops of ribbon on a wreath.

In my version of the Afterlife, the winters are mild, the skies are not cloudy, and the analemma dances overhead all day.

Deities of the Sun—God and Goddess

In the same way that some people call every cat "she" and every dog "he," we often call the Moon "she" and the Sun "he." This works well—except when it doesn't.

In ages past, the Winter Solstice was used to mark the birthday celebrations of the gods Attis (Asia Minor), Dionysus (Thracian-Greek), Osiris (Egyptian), Baal (Assyrian), Zeus (Greek), and Cuchulain (Irish).[40] Long before December 25th was officially designated as Christmas to celebrate the birth of Jesus Christ, it was known as the first day of northerly solar motion after the solstice's standstill. As such it was widely celebrated—as the Roman *bruma* or "the shortest day," and as the god Mithra's *Dies Natalis Solis Invicti,* the Birth of the Unconquered Sun. The annual rebirth of the Norse god Frey was celebrated at this time as *Yule,* which means "wheel."[41]

Other male solar deities include Helios (Greek), falcon-headed Re (Egyptian), Surya (Hindu), Tonatiuh (Aztecs), and Inti (Incan), whose face graces the flags of Argentina and Uruguay. At Newgrange in Ireland, the Winter Solstice Sun was viewed as the male principal penetrating into the female Earth.[42]

The Sun's energy can be perceived as masculine in another way as well. Many standing stones are roughly phallic in shape, and they become shadow-casting gnomons when the Sun strikes them. Their shadows are longest—most virile—at Winter Solstice.

Figure 48. The flattened Möbius strip, and the recycling symbol.

Flipping the gender stereotype, plenty of cultures have perceived the Sun as female. For the early Norse, the goddess Sól drove her Sun chariot through the skies by day and illuminated the underworld realm of Hel by night. Among the Germans, goddess Sunna traveled the daytime sky; her departure to sit on her unseen throne each night is the origin of our term *sunset.*[43] The Sun was Saule to the Baltic people, Solntse to the Slavs, Grian to the Celts, and Arinna to the ancient Hittites—all female divinities.

In Asia Minor, the Sun was the goddess Kubaba, who gradually shifted into the Greco-Roman Cybele, whose worship stretched throughout Europe. Kubaba is often depicted astride a lion; Cybele rides in a chariot pulled by lions. Both are also magnificently portrayed seated on a throne with giant cats on each side.

Figure 49. Sun-crowned and lion-headed Sekhmet, Egyptian Sun goddess.

In the Egyptian pantheon, the Sun was sometimes personified as the Sun-crowned and lion-headed goddess Sekhmet. Rays like petals or sunbeams flare back from her face (see figure 49). When appeased, Sekhmet removed fever and illness; otherwise, her overheated breath created deserts.

Why are all these deities associated with lions? Perhaps because from roughly 4065 to 1450 BCE, the long years during which many of these deities emerged, the Summer Solstice Sun was in line with Leo.

In Japan, the Shinto Sun goddess Amaterasu is especially prominent. She's considered the ancestress of the Japanese ruling family and is represented by the red circle on the Japanese flag.

Bringing these female and male strands of lore together, Christmas Eve—known of old as *Modranect,* or "Mother's Night"—was considered an even greater festival than Christmas Day itself.[44] Mother's Night emphasized long gestation and the act of giving birth, while the solstice itself emphasized that which was born. The night and its dawn were treated as companion holidays—a balance of female and male energies that brought back the light.

Angles

To demonstrate a few ideas about the Sun's motion and how our early ancestors used the information, we'll use the locations of several specific Northern Hemisphere sites: Rösaring (Sweden), at 59°30' N; Stonehenge (England), at 51°11' N; Chaco Canyon (New Mexico, USA), at 36°3' N; and Chichen Itza (Mexico), at 20°40' N.

Take a look at figure 50, which is drawn as if you, the viewer, are standing at each site and watching a solstice sunrise (right) or sunset (left). You can see here the wide variation in angles from site to site. The Sun is directly above the solstice latitudes of 23° 26' N or 23° 26' S, but what *you* see depends on your earthly vantage point, with wider angles occurring between solstice points as you move north, and narrower angles occurring as you move south.

We find Xs, zigzags, and diamond and lozenge shapes used worldwide as decorative motifs (see figure 51). Did this angled artistic element also function as an expression of location, indicating where our ancestors came from?[45]

Let's look at several variations. From top to bottom in figure 52, we see a basic lozenge shape (a), and then see the same angles arranged as an X (b)—closed or open shapes. Moreover, you can make each angle into a separate significant object, like the rune *Gyfu* (meaning "gift" and "reciprocity"), or a drafting compass and carpenter's square, like the symbol of the Freemasons (c).

Moving down the column in figure 52, the angles become the rune *Jera* (d), meaning both "year" and "season"—in the sense of having a full cycle of time, making Jera's place as the twelfth rune very fitting. Other meanings—harvest, harmony, orderly progression, fruitful results—are also carried by Jera, all compatible with an expression of a full year, coming full circle.

Arranged with one joined to the other, the angles become *Sowilo* (e), which most simply means "Sun." In old Scandinavia, the Sun was perceived as female,

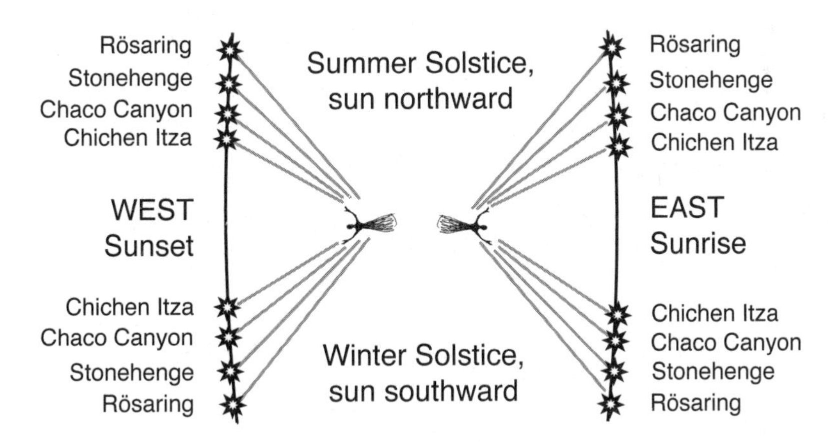

Figure 50. Comparative solstice positions and angles.

and this rune relates to the goddess Sól.[46] Look at the zigzag of this rune and imagine it extended as a linear pattern like rickrack, perhaps as an expression of latitude or time.

The same angles overlapped become the rune *Inguz* (f). Ing was a fertility god and consort of nurturing Earth-goddess Nerthus, and His rune was symbolic of light, whether a hearth fire or a signal fire. Another fertility god, Frey—also known as Ingvi-Frey—was honored as a god of sunshine and happiness in marriage. Frey's birth was celebrated at the Winter Solstice. The Inguz rune can work as an illustration of sexual union—PG-rated, but imagination-sparking.

The word *rune* means "secrets" and "whisper," and, true to those ideas, runes often function subtly on several levels and with multiple layers of meaning.

The final symbol in figure 52 is *Awen* (g). In Welsh, *awen* means both "muse" and "rein," and in modern Druidry, the word refers to "inspiration." Awen can symbolize a variety of triplicities, including body-mind-spirit, earth-sea-air, and

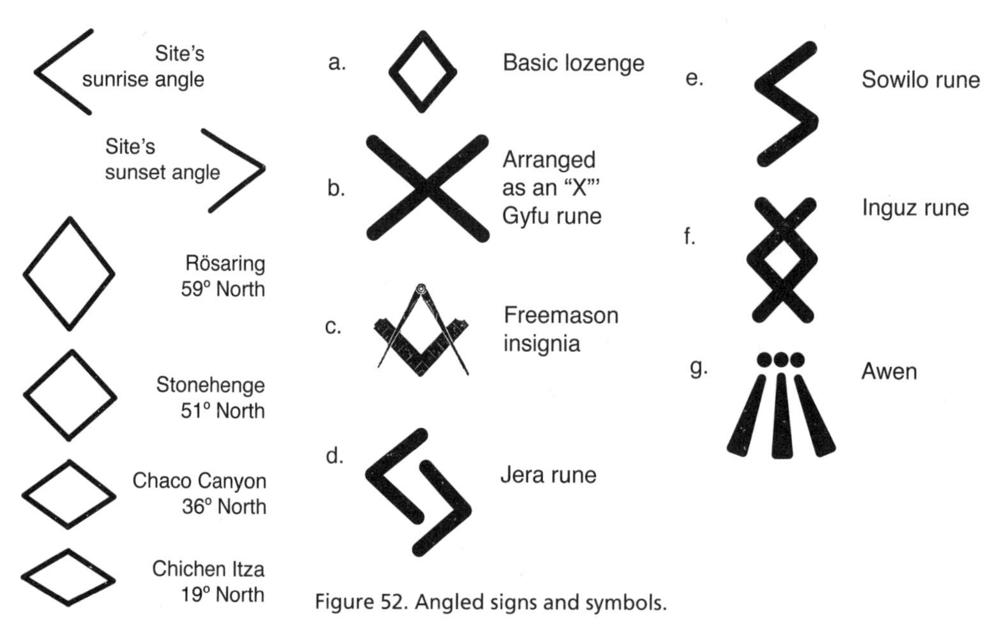

Figure 51. Lozenge and diamond shapes, differing by latitude.

Figure 52. Angled signs and symbols.

love-wisdom-truth. It also expresses the Sun's three positions—one line for each solstice and the central line shared by the equinoxes.

When we greet the Sun, the Moon, and the four quarters, many of us use an arms-up "invoking" gesture that echoes these V-shaped angles. For me, the position tends to be a gesture of both openness and embrace, a moment to breathe deeply, drink in the wonders of whatever direction I'm facing, and attune my attention, my intentions, and my focus. Maybe our stretch to the sky is also an ancestral memory of our earlier locations. Food for thought the next time you greet the rising Sun and invoke the energies of the new day.

The Sky Expressed in Architecture

People throughout time have expressed their understanding of planetary motion in their architecture. While celestial observations obviously played a key role—these places were *designed* to showcase the specific details of the planets—many of these were places that brought people together and evoked the mysteries of spirit. These were also places of awe and wonder and connection with the Beyond.

There are thousands of sites, with more discovered each year, and their imagery ranges from straightforward to stunningly complex. Your location dictates all. You can't just figure out one master plan and then sail off to build identical monuments elsewhere, like a string of franchised burger joints. For example, alignments that work for Stonehenge at 51° N latitude are wrong farther north or south. So even if builders brought a wealth of sky knowledge with them to a site, they still had to fine-tune the details once they arrived.

Here's a small sampling.

Rösaring, Sweden (roughly 59° 30′ N, 17° 30′ E; c. 1700 BCE–1200 CE)

A fifteen-circuit stone labyrinth gives this site its name. At Rösaring's location, summer has eighteen hours of daylight; winter has only six.[47] The Sun is barely 7° above the horizon at the Winter Solstice. A unique ceremonial roadway is positioned to take best advantage of a near-exact north-south alignment. The road is short (1774 feet/540 meters); at its southern end, there is a low, flat-topped mound and a panoramic view over Lake Mälaren.

Some speculate that this may have been a ceremonial processional route for a wagon carrying a sculpture of the Earth goddess Nerthus, as described by Roman author Tacitus in *Germania*, c. 96 CE.

Edged by tall pine trees then as now, the road is dark at the Winter Solstice until noon. Then the low Sun shines into this tree-lined "canyon" from the south, lighting first the western side, then shining straight up the avenue, then lighting just the eastern wall of trees (see figure 53). Thirty minutes later, the avenue is dark again. The sunlight remains directly centered for about fifteen minutes, just enough time to traverse the full length and reach the mound, as if walking into the Sun's orb along a tree-walled temple of snow-spangled light.

Stonehenge, England (51° 11′ N, 1° 49′ W; c. 3150–1950 BCE)

Stonehenge is unique for its trilithons, three-stone archways formed from two massive uprights joined by a cross bar or *lintel* across the top (see figure 54). The Summer Solstice sunrise is marked here with a sightline and trilithon aimed out to the Heel Stone. The tallest trilithon stands 16 feet tall (4.9 meters). Rising and setting positions of equinoxes and solstices are marked, as are lunar extremes (treated in chapter 4) and possibly eclipses.

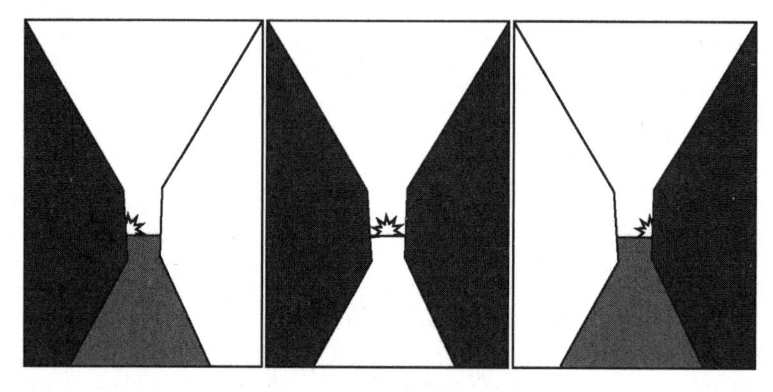

Figure 53. Shifting sunlight on the Rösaring roadway.

Figure 54. Stonehenge and a rising Summer Solstice Sun over the Heel Stone.

Goseck Circle, Germany (51° 12′ N, 11° 51′ E; c. 4900 BCE)

At almost exactly the same latitude as Stonehenge, Goseck is about a mile (about 1.85 km) farther north. Rediscovered in 1991, this Neolithic ditch-ring and restored wooden-palisade circle points out the Winter Solstice sunrise and sunset through gateways in the palisade fence. Goseck considerably precedes Stonehenge, and perhaps led to it. (Note: Of two hundred ditch-ring sites known in Germany, Austria, and Croatia, only about twenty have been investigated.[48])

Newgrange, Boyne Valley, Ireland (53° 41′ N, 6° 28′ W; 3100–2900 BCE)

Newgrange is a huge passage tomb with a grass-turfed roof. Much of its outer oval wall is faced with gleaming white quartz.[49]

It channels sunlight via a "light box," a transom-like window over the main door that directs the light to the back wall of a 59-foot-long (18 meters) upward-slanting passageway (see figure 55). The light box is roughly 39 inches (almost 1 meter) wide and 10 inches (25.5 cm) high.

Figure 55. The Newgrange "light box" and Winter Solstice sunlight.

Chaco Canyon, New Mexico, USA (36° 3′ N, 107° 57′ W; c. 850–1140 CE)

Situated in a desert, Chaco Canyon contains nearly four thousand archaeological sites, the best known of which is the "Sun dagger" feature on Fajada Butte. There, three large thin slabs of stone lean against a cliff edge, allowing narrow bands of sunlight through to illuminate portions of the wall behind them, on which spirals are carved (see figure 56). At the solstices and equinoxes, "Sun daggers" of noontime light pierce or bracket the spirals.[50]

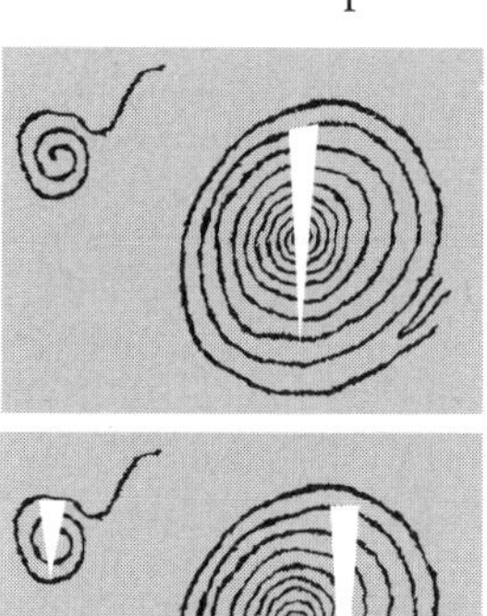

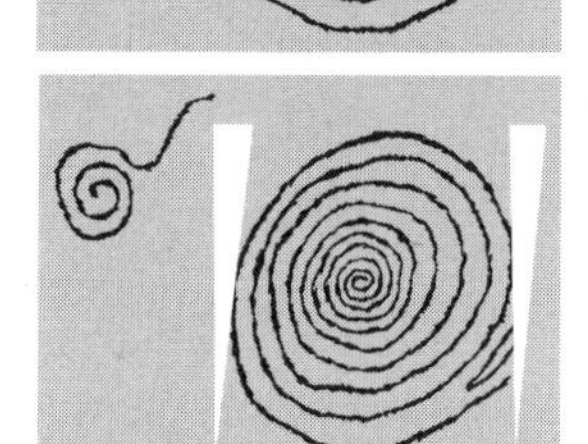

Figure 56. Fajada Butte's "Sun daggers." From top to bottom, at Summer Solstice, at each of the equinoxes, and at Winter Solstice.

A small petroglyph elsewhere on the butte echoes Chaco's main building, Pueblo Bonito (see figure 57). This huge structure was originally four or five stories tall, with hundreds of rooms and dozens of kivas, or below-ground ceremonial chambers. In figures 57 and 58, each circle represents a kiva; the key north-south wall through the open courtyard is shown, like the petroglyph's arrow.

We use that north-south wall to find the Sun at solar noon, which is always due south at its zenith, although the Sun's declination varies throughout the year. Shadows on each side of the courtyard's low walls mark the Sun's motion daily (see figure 58). When the Sun reaches its zenith, the shadow vanishes, marking time by its absence. Any true north-south wall does the same trick at solar noon.

Chichen Itza, Yucatan, Mexico (20° 40′ N, 88° 34′ W; c. 600–1000 CE)

Chichen Itza is one of thousands of city/sacred-center sites established by the sky-savvy Maya. Its great Temple of Kukulcán is a nine-terraced step pyramid capped with a small temple (see figure 59). Stairways containing ninety-one steps each lead up the structure's four steep sides to a platform, which acts as the final shared step (91 steps x 4 sides = 364; the final shared step brings the total to 365, the number of days in a year). The edges of the northerly facing stairway end in huge serpent heads that rest on the ground with mouths agape.

The temple is aligned aslant, 19° east of celestial north. The afternoon Spring Equinox Sun casts shadows while lighting only the right-hand snake head and a wavy line along the stairway's edge. As the Sun moves west, a giant sunlight snake appears to ripple down the pyramid's shadowed terrace.

Serpent Mound, Ohio, USA (39°1′ N, 83°25′ W; c. 1070 CE)

Figure 60 shows another serpent, an effigy mound called Serpent Mound. It begins in a tight coil, opens into undulating curves, and resolves into a wide-open mouth around a huge oval earthen "egg." The effigy stretches 1,348 feet (410.8 meters) along the ground and, after centuries of weathering, averages three to five feet (.91 to 1.52 meters) in height. This is the New World's largest known effigy earthwork.

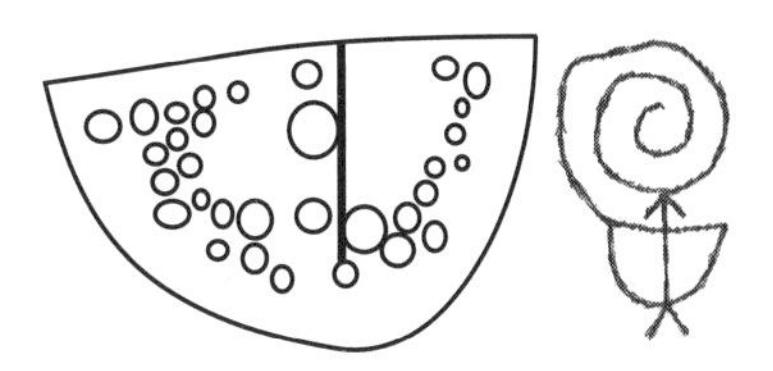

Figure 57. Pueblo Bonito (left) looking south, and the Fajada Butte petroglyph.

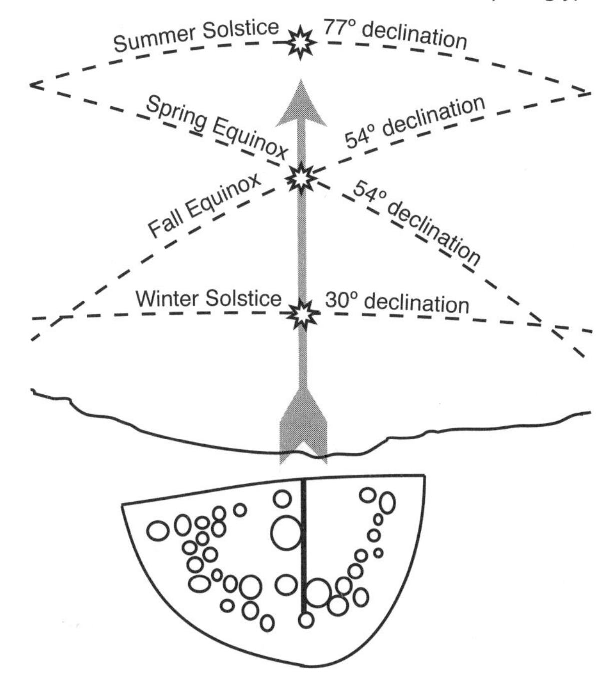

Figure 58. Pueblo Bonito's solar-noon alignment.

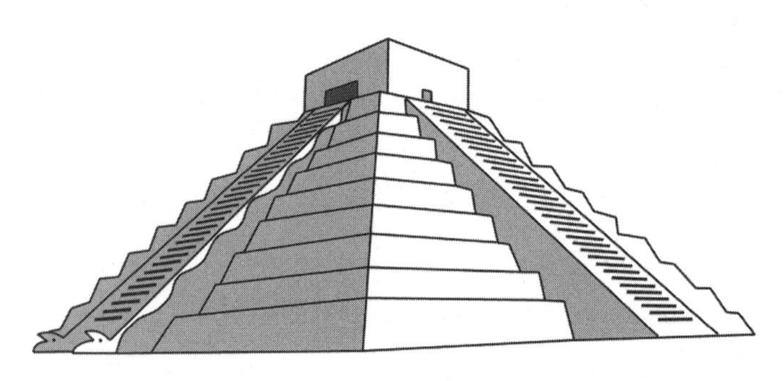

Figure 59. The Temple of Kukulcán at the Spring Equinox with serpent-heads at lower left..

Many American states remain rich in mounds, even after untold numbers were destroyed as impediments to progress.

The serpent's head and the centerline through its egg act as a pointer to the Summer Solstice sunset.

Mounds and other earthen forms stretch into the current era. "Earthworks" emerged in contemporary art beginning in the 1960s, with artists striving to move away from art as an object or marketable commodity. Viewing such artwork is experiential, as the works tend to be large, sculptural, and out of doors. The best-known modern American earthwork, although seldom defined as such, is the Vietnam Veterans Memorial in Washington, D.C. The sunken black marble walls align to the Lincoln Memorial and the Washington Monument rather than to cardinal directions or to the Sun, but the memorial embodies the spiritual and emotional potency inherent at ancient sites, amplified here by our personal associations. Significantly, the memorial's architect, Maya Lin, grew up in Ohio and, as a child, made school trips to various ancient mound sites, which she cites as an enduring influence.

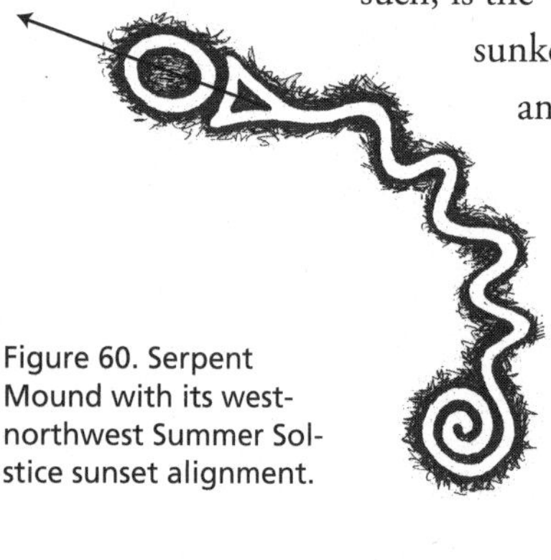

Figure 60. Serpent Mound with its west-northwest Summer Solstice sunset alignment.

Jantar Mantar, Delhi, India (26° 55′N, 75° 49′E; c. 1727–1734 CE)

Mughal ruler Maharaja Sawai Jai Singh II (1688–1743) built observatories called Jantar Mantar in five of his cities. Like a giant's geometric playthings, the instruments are the size of buildings; the largest is seven stories tall. In things astronomical, greater size means more stability and greater accuracy. On the instrument shown in figure 61, time is marked by a shadow moving along the calibrating scale: 1 millimeter for each second, half the width of a hand each minute. Here, you actually *observe* the passage of time.[51]

Jantar Mantar means "instrument of calculation," but can also be translated as "magical instrument" and "magic sign."

Magic—*real* magic, a connection back to spirit—is at the heart of all this monumental mark-making. The magic lies in how we feel in the midst of it. Simply marking a solstice or equinox position requires just two large rocks to establish sightlines: stand here, look there. Why do more? Sunrises and sunsets have their own grandeur and don't need our building projects to make them better.

But *we* need *them*—these constructions, these temples. They express our understanding of the Sun's journey through time, and the recurring cycles that carry us along in their changes. They express our cultural connections to the seasons, to our goddesses and gods. Perhaps—like a giant signal—they are meant to let the heavens know: "We heed you." Like every shrine, they give us a location

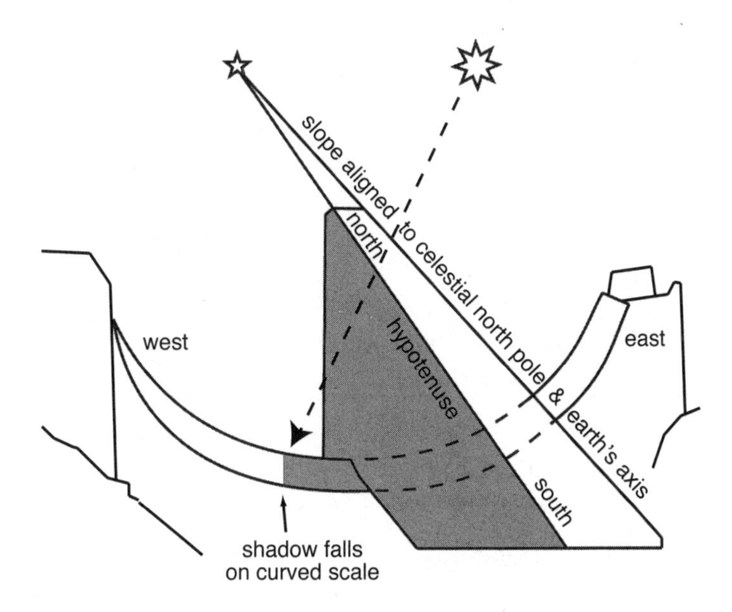

Figure 61. How the Jantar Mantar's huge Samrat Yantra works.

to experience whatever they mark, alone and with our communities. They help us step into the living experience of time.

Learn by Doing

1. **Use shadow time:** Our sense of time is significantly different than that of our ancestors. Try to attune yourself to the Sun's motion as a means of gauging time. Shadows are an expression of this, of course, so heighten your awareness of them as well, or try working just with shadows.

2. **Make a sundial:** Stand a gnomon stick up vertically somewhere that it will get year-round sunlight, at least at one specific time of the day. Make sure that there's flat space around the stick for shadows to fall, and space that can be left undisturbed. About every ten days, *always at exactly the same time,* make a mark on the ground at the tip of the stick's shadow. Whatever time you pick, stay absolutely consistent and ignore the shifts of Daylight Saving Time. You'll see the analemma pattern quickly, long before an entire year has passed.

3. **Be your own gnomon:** Photograph your own shadow at high noon on your birthday as an indirect form of self-portraiture, and as a means of exploring and personalizing time. Or mark this out along the ground with stones or flowers (and a friend's assistance). If you were born during the daylight hours, you can mark your shadow at your exact birth-time.

4. **Make a Möbius strip:** Cut a strip of paper about 1 inch wide and 11 inches long. Bring the narrow ends together to form a circle, but before joining them with tape or a staple, flip one end over so the "front" of one end meets the "back" of the other in a paper loop with a twist in it. Draw one line down the long middle of the strip without lifting the pencil from the paper. You'll return to your starting point, but your pencil line will have traveled along both sides of the paper.

5. **Honor solstices and equinoxes:** Mark the solstices and equinoxes with personal ceremonies, greeting the new season as you bid farewell to the one

that's ending. Or work in a more graphic way, observing sunrise or sunset on each solstice and equinox. What landmarks in the east or west can you use as markers for the Sun's motion? What "landmarks" in your own life repeat or shift through the course of the year? Marking the seasons honors the cycles and changes in our own lives so that we stop being just observers. Through land and sky, we step into the flow of the year's cycle and become, season to season, part of this larger rhythm.

Chapter 3

The Precession of the Equinoxes

Although it has little to do with how we look at stars and planets now—one night at a time—the precession of the equinoxes has plenty to do with how our distant ancestors saw the sky and expressed what they found there. In a way, it delineates how our view differs from theirs.

Most of us first heard about the precession of the equinoxes in reference to the Age of Aquarius. From there, the interpretations branch off in all directions.

Our world and solar system are, happily, calibrated to such a fine degree that many things recur dependably, year after year. But that calibration isn't absolutely exact, and Earth's orientation within the well-ordered universe slowly shifts with time. Like a clock losing one minute of time each year, Earth loses 1° of calibration every seventy-two years. That's gradual, but it adds up. Recognition of the precession of the equinoxes, as this slippage is called, is credited to the Greek astronomer Hipparchus (c. 190–120 BCE), whose work benefited from that of earlier Babylonian astronomers. Many early cultures restarted their annual calendars with the Spring Equinox, so this phenomenon is named for, and tracked through, that solar holiday.

When we celebrate the Spring Equinox around March 20 each year, the Sun is indeed halfway between its two solstice points. But as for the Sun entering the first degree of Aries on that day—nope. That hasn't been correct since around 50 BCE. After that rough date, the Sun's Vernal Equinox point wasn't in Aries at all;

instead, it was moving back through Pisces. This shift is called *precession* because the Sun is gradually lagging back through each *preceding* sign for its Spring Equinox location. This is similar to a slow-running clock that reads 11:42 a.m., even though it's really noon. Nowadays, although the halfway-between-solstices event of the Vernal Equinox still occurs around March 20, the Sun is in Pisces on that date and doesn't enter Aries until mid-April.

The precession happens because of several factors working in combination:

1. The Earth isn't a perfect sphere: it's an oblate spheroid.
2. The Sun, Moon, and planets exert gravitational forces that squash and pull the Earth.
3. The Earth is tilted on its axis about 23.5°—without which we wouldn't have seasons (see figure 62).
4. The Earth's orbital motion isn't *exactly* synchronized to the rhythm and timing of the heavens.

Damn it, Jim, I'm an artist, not a rocket scientist. To comprehend how these factors combine, picture yourself spinning counterclockwise with your arms outstretched and a nicely pointed witch hat on your head. Try to keep the hat's point—and your whole body—aligned to a specific star overhead. It's difficult because you're on a hillside rather than on flat ground (tilt!) and are holding a grapefruit in one hand, which throws off your balance (the oblate spheroid/gravity effect). Just to keep things interesting, there's a ram grazing nearby: please point at his head each time you spin around (the equinoctial rhythm).

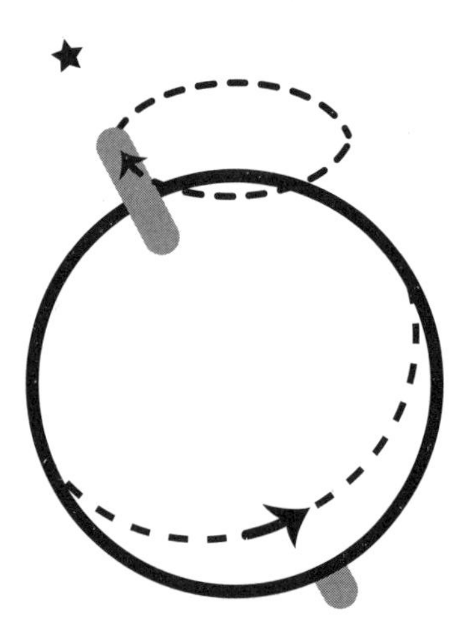

Figure 62. The tilt of the Earth on its axis, with Polaris, our current North Star.

Although your pointing rhythm is great, you're not spinning quite as fast as you'd like, so lately you've been pointing at some fishes off to the ram's right. Eventually, you'll point at the water-spilling fellow standing to the right of the fishes. Voilà! Precession of the equinoxes.

So the Earth's orientation in the heavens shifts subtly over time, which causes two things: the constellations shift in relation to our solstices and equinoxes; and, eventually, the Earth has a new North Star. Let's look first at the precession and then at the North Star's shift.

Precession

The idea of "ages"—as in the Age of Aquarius—is based on the precession of the equinoxes. The explanation goes something like this.

The Spring Equinox point shifts over time (true), coming gradually into line with one constellation after another (true) while moving *backward* through the zodiac signs (true), which brings about worldwide changes and shifts in human consciousness (open to interpretation).

Our distant ancestors saw a different springtime sky and responded accordingly. For example, during the long years when the Spring Equinox point was in line with the constellation Taurus, many cultures portrayed cows and bulls prominently in their art, sometimes worshipping deities with cow/bull/oxen attributes. As the Spring Equinox point shifted to align with Aries, more ram imagery appeared. As the Spring Equinox moved into alignment with Pisces, Christianity emerged, with its fish-related symbols and descriptions of Jesus as "a fisher of men." Cause or coincidence? Did they envision bulls and rams in the sky because they were busy domesticating these animals, or did the shifting equinox point *cause* their changes in worship and culture? That's a matter of opinion.

The Age of Aquarius has been envisioned as a time of expanding consciousness and great spiritual leaps of comprehension and understanding. Until then, we are living in the Age of Pisces, which just means that, on the morning of the Spring Equinox, the Sun rises with the constellation Pisces behind it. The vernal equinox point is currently about 8° in from the official edge of Pisces (which isn't the same thing as Pisces 8°)[52] and moving gradually closer to that Pisces-Aquarius "border."

The entire cycle of the precession of the equinoxes, from 0° Aries backward through the zodiac and finally returning to 0° Aries, takes about 25,800 years.[53]

If we divide 25,800 years by the twelve zodiacal constellations, we get twelve periods of 2150 years each. Great! If each age is 2,150 years long, that means that, if the Sun rose in line with Aries for the first time on the Spring Equinox around 2200 BCE (give or take a century or so), it must have risen in line with Pisces for the first time on the Spring Equinox around 50 BCE, and will finally rise in line with Aquarius for the first time on the Spring Equinox around 2100 CE.

Working with twelve constellations, at 30° each and a steady 2150 years per sign, each constellation's equal time as the home of the Spring Equinox looks like this:

Aquarius	c. 23,700–21,550 BCE
Capricorn	c. 21,550–19,400 BCE
Sagittarius	c. 19,400–17,250 BCE
Scorpio	c. 17,250–15,100 BCE
Libra	c. 15,100–12,950 BCE
Virgo	c. 12,950–10,800 BCE
Leo	c. 10,800–8650 BCE
Cancer	c. 8650–6500 BCE
Gemini	c. 6500–4350 BCE
Taurus	c. 4350–2200 BCE
Aries	c. 2200–50 BCE
Pisces	c. 50 BCE–2100 CE
Aquarius	c. 2100–4250 CE

This works great on paper (see figures 63 and [on page 81] 65). If only the stars would oblige. But they can't, because the zodiac's constellations vary in size. Why bother dividing 25,800 years equally by 12 when these constellations aren't equal? Check the diagrams of the zodiacal constellations, figures 21, 24, 28, and 31, back in chapter 1. No matter where those official vertical or horizontal lines cut through, you'll see that some constellations stretch wide from east to west along the ecliptic, while others occupy just a narrow space there.

Figure 63. The zodiac circle divided into equal 30° sections.

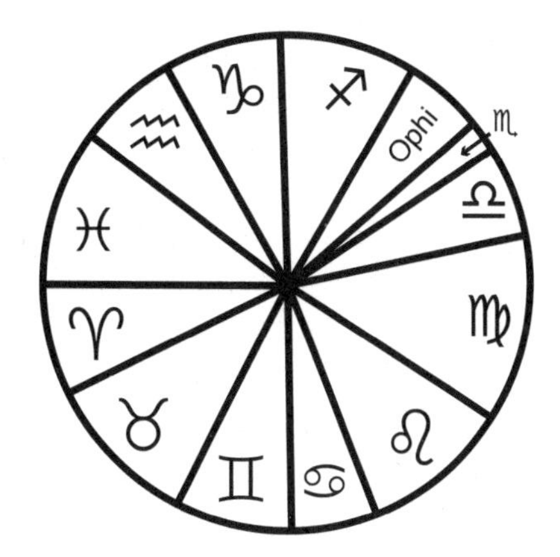

Figure 64. The zodiac wheel showing all the ecliptical constellations based on their more accurate "official" dimensions.

Instead of twelve zodiacal constellations each occupying 30° of sky, when we measure along the ecliptic with these official modern borders, the real width of every constellation along the ecliptic looks more like this:

Aries	24°
Taurus	37°
Gemini	28°
Cancer	20°
Leo	36°
Virgo	44°
Libra	23°
Scorpio	6°
Ophiuchus, the Serpent-Bearer	19°
Sagittarius	34°
Capricorn	28°
Aquarius	24°
Pisces	37°

Applying these modern boundary lines between constellations back through time, the zodiac wheel looks more like that shown in figure 64, and the precession dates look more like this (see figure 66):

Aquarius	c. 23,440–21,600 BCE (1840 years)
Capricorn	c. 21,600–19,600 BCE (2000 years)
Sagittarius	c. 19,600–17,250 BCE (2350 years)
Ophiuchus	c. 17,250–15,900 BCE (1350 years)
Scorpio	c. 15,900–15,295 BCE (605 years)
Libra	c. 15,295–13,765 BCE (1530 years)
Virgo	c. 13,765–10,600 BCE (3165 years)
Leo	c. 10,600–8000 BCE (2600 years)
Cancer	c. 8000–6560 BCE (1440 years)
Gemini	c. 6560–4530 BCE (2030 years)
Taurus	c. 4530–1900 BCE (2630 years)
Aries	c. 1900–50 BCE (1850 years)
Pisces	c. 50 BCE–2590 CE (2640 years)
Into Aquarius	c. 2590–4325 CE (1735 years)

Look at the disparity in these time periods! Based on the 1930 IAU boundaries, our Vernal Equinox and the symbolic rebirth of Spring won't truly precede into Aquarius until around 2590 CE. Remember, these redrawn boundaries aren't the source of the discrepancy. The zodiac signs were *never* equal in size. The constellations are just a human construct of connect-the-dots images superimposed on the wild and vast starry sky, and nobody paced off 30° slices—color inside the lines!—before they began making pictures out of the stars.

This fact scrambles a lot of well-established concepts. For example, during the long years that the Spring Equinox aligned with any portion of wide Taurus, the Autumn Equinox occupied four different signs—from Sagittarius, to Ophiuchus, to skinny Scorpio, and finally into Libra. Moreover, the Autumn

Figure 65. Precession on the astrological 30° zodiac wheel. Top: 4530 BCE; Sun precesses into Taurus. Bottom: 2200 BCE; Sun precesses into Aries.

Figure 66. Precession of the equinoxes on the zodiac wheel according to the new official boundaries. Top: 4530 BCE; Spring Equinox precesses into Taurus. Bottom: 1900 BCE; Spring Equinox precesses into Aries.

Equinox's shift into Libra came *before* the Spring Equinox moved into Aries, rather than simultaneously as shown in figure 65. Compare the standard version in figure 65 and the more accurate version in figure 66 to see this contrast.

This leads us to a third, more naturalistic, way of looking at the zodiac constellations, based on their actual appearance in the sky. Along the ecliptic, this includes many vague periods when the Spring Equinox seems to hover between two constellations, in the empty-looking space between them. Or constellations may physically overlap—remember that the balance-pan stars of Libra's scales are named for Scorpio's claws (see figures 67 and 68).

Figure 67. The zodiac circle without lines, with the size of the glyphs expressing each sign's relative size or distinctiveness.

We measure and calibrate and officially define. Our ancestors did their own versions of those things, but they also observed. Without strict concepts of 30° signs (astrology) and official boundaries (modern astronomy), they simply eyeballed it.

How do we perceive the starry backdrop? Are the Moon, the Sun, the planets, or the solstice or equinox points closer to this constellation or that one? When the sky's spiritual power outweighs the need for precision, the outcome shifts organically, as shown in figure 69. These are the equinox and solstice positions of our current era, what we see.

Despite a certain human desire for quantifiable detail and exactitude, my star gazing isn't an

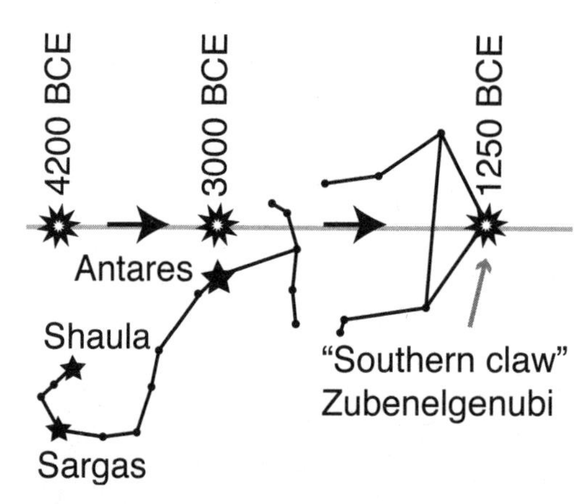

FIGURE 68. A no-borders, flexible view of Scorpio's duration as the location of the Autumn Equinox.

exact science. I'm navigating my spiritual life—not a space ship—and my signposts are the beliefs, symbols, *feelings,* and other intangibles that hold significance for me.

Where does this leave us? With the power of the experience. For me, as for many of us, the solstices and equinoxes resonate deeply as a recognition of time, of recurring cycles, and of the spiritually resonant movement of the Self through Time. This empowered personal awareness trumps the need for navigator-style precision.

That said, navigating *reality* still ranks high, and we aren't in the Aquarian Age quite yet. That doesn't mean we aren't experiencing heightened awareness and shifts in consciousness. It's just that some of these realizations look distinctly Piscean. Fishes, water-element sign—hello, water issues! Amid oil spills, pollution, tsunamis, and aquifer depletion, awareness of water is a wave that is carrying us out of Pisces and on toward the Aquarian Age.

Figure 69. The border-free zodiac wheel showing the current solstice/equinox points.

Göbekli Tepe—The First Temple Ever? (Southern Turkey, 37°13' N, 38°55' E; c. 9500–8000 BCE)

A few years ago, it would have seemed wildly speculative to look back as far as 9500 BCE for validation of sky watching, but Göbekli Tepe has changed all that. In fact, this huge site in southern Turkey may actually date from as early as 11,500 BCE. Then, in a massive undertaking around 8000 BCE, it was *intentionally* filled in—buried—a baffling action that hid the site and helped preserve it. Even with the more conservative 9500 BCE dating, Göbekli Tepe is farther away in time from Stonehenge than Stonehenge is from us.

Göbekli Tepe is believed to have been a gathering place for nomadic hunter-gathers who eventually formalized their meeting ground at the site. The shaped and positioned pillars are carved with a range of animals, including asses, boars, bulls, foxes, gazelles, insects, lions, scorpions, snakes, vultures, and water fowl, as well as an occasional human. There are many individual temples here. Four have been uncovered so far, and head archaeologist Klaus Schmidt estimates another fifteen to twenty are still buried.[54] It is also important to note what *hasn't* been found—dwellings. So far, there's nothing like a home-sized foundation, fire-pit, or midden that would point to Göbekli Tepe as being part of a village.

Figure 70. Göbekli Tepe's bull.

Looking just for the zodiacal animals (as we identify them) on the list of carvings at Göbekli Tepe, we find large felines, scorpions, and bulls. This is interesting, since Scorpio, Taurus, and Leo were Summer and Winter Solstice and Spring Equinox positions respectively during Göbekli Tepe's era. On one great pillar, a carved bull seems to carry the Sun between its horns (see figure 70). From roughly 11,065 to 8500 BCE, the Winter Solstice occurred in the constellation of Taurus, so perhaps a bull carried the Sun back to these people from its Winter Solstice southern extreme.

Göbekli Tepe is rewriting much of what we previously believed about early civilization, early agriculture, and possible ritual practices. Apparently humans built ritual centers *before* they settled down to grow crops, not after. The site's deliberate burial around 8000 BCE roughly coincided with the precession's inevitable shift of the Vernal Equinox away from Leo and into the next sign, faint Cancer (see figure 71). This timing suggests an explanation for the site's abandonment. Simply put, it was no longer accurate. Perhaps they felt their huge, elaborate clock had betrayed all their hard work.

Here's another mystery. Throughout the world, nomadic people are geniuses of *portable* culture. By necessity, their ingenuity is focused on crafting beautiful, utilitarian items that travel well. Huge stones are the antithesis of portability. They *don't* travel well. Göbekli Tepe's pillars weigh ten, twenty, and even fifty tons (9.07, 18.14, and 45.36 tonnes) each. How, when, and why did pre-agricultural nomads master the art of moving, assembling, and carving stone on this colossal scale?

About 95 percent of the Göbekli Tepe site is still buried, unseen for ten thousand years. There is plenty yet to be discovered, and many answers are likely to emerge. The site's astronomical uses, if any, have yet to be deeply explored. Judging from the sky-oriented themes that have called forth massive human endeavor elsewhere, however, it's reasonable to suspect that there *will* be clear astronomical components to Göbekli Tepe as well.

The Great Serpent and Thuban—An Earlier North Star

Remember, a shift in the Spring Equinox point isn't the only effect that the precession has had on our skies. In the opening chapter, we met our current North Star, Polaris, and saw that the circumpolar stars never set. Look again at the spinning globe in figure 62, and you'll see a star above it, in line with the Earth's axis, or pole. That's our current North Star, Polaris.

However, as the loop alongside the pole in that illustration is meant to show, through the ages the Earth's pole makes its own slow circuit of the never-setting circumpolar sky. The precession of the equinoxes very gradually alters the orientation of the Earth's pole. We currently enjoy a bright, nearly exact North Star. Over the long history of sky-watching humans, however, there have been years—even centuries—when the Earth's pole pointed to an area with no stars, or only very faint stars, in that true-north position. Nonetheless, our travel-minded ancestors got around just fine without an exact North Star.

Figure 71. Göbekli Tepe's range of ages—from 9500 BCE (top) to 8000 BCE (bottom).

During a significant portion of ancient times, however, a star called Thuban was the stable North Star, occupying that position from around 3400 to 2300 BCE, most precisely around 2800 BCE. Thuban is in the constellation Draco, the Dragon, which slithers between the Bears in the northern heavens, looking more like a snake than a dragon to modern eyes.

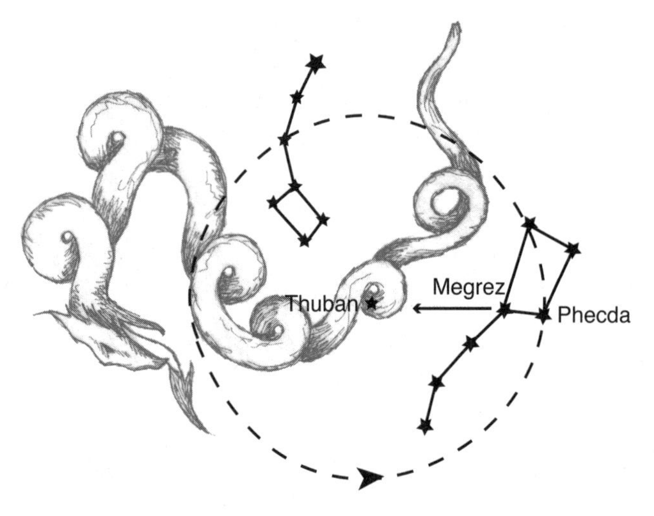

Figure 72. Draco and Thuban.

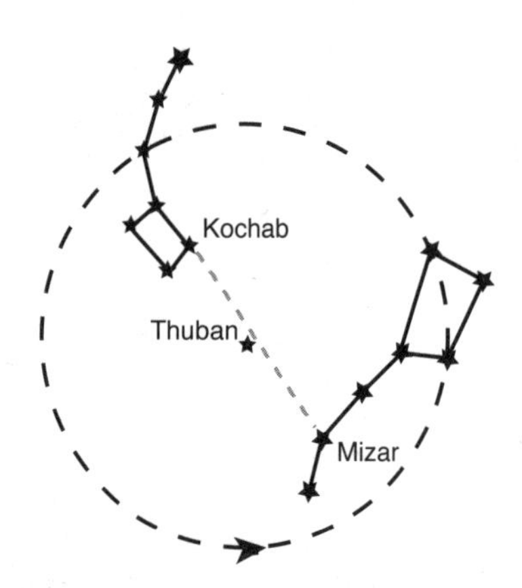

Figure 73. Kochab and Mizar, bracketing Thuban.

Thuban's attraction was its position, not its brightness—it's pretty faint—but you can locate it in the same way that our ancestors could. Use the stars at the other end of the Big Dipper's pan, Phecda ("the thigh" of the Great Bear) and Megrez ("the root of the tail"), both along the sides of the "dipper" nearest the handle (see figure 72).

Even when Thuban wasn't well centered, it was still the best thing available. Astronomical texts from Mesopotamia (c. 2300 BCE) give Thuban several exalted titles—*Tir-An-na*, the Life of Heaven; *Dayan Same,* the Judge of Heaven; *Dayan Sidi,* the Favorable Judge—even though Thuban was then about 2° away from true celestial north.[55] Remember, that's only the width of one finger held at arm's length (see figure 11).

The Great Pyramid of Khufu (Giza, Egypt, 29°58' N, 31°49' E; c. 2580–2560 BCE)

This first and largest pyramid at Giza contains a symbolic alignment to Thuban, out among the circumpolar stars. Angling down into the Great Pyramid from its north-facing side is an "airshaft," so-called only because of its petite nine-by-nine-inch size (23 by 23 cm). The shaft points out to the north sky at a 31° angle, aimed at the spot then occupied by Thuban at its culmination. Thuban was less than 1° off true north in those days, so the star was actually tracing a tiny circle around the north axis. Thuban culminated—reached its highest point—when it swung around to sit above true north, 31° above the horizon. Thuban was then symbolically connected with the King's Chamber by means of the airshaft.

Pyramids were architecture of the spirit, and Khufu's airshaft was not an observatory so much as a symbolic marker. For the Egyptians, the circumpolar northern stars symbolized Eternity, the dwelling place of the pharaoh's ancestors. Never rising, never setting, like a stable vanishing point on the celestial road, these stars directed the pharaoh's spirit into Eternity.[56]

In fact, the Egyptians called the circumpolar stars the Indestructibles, the Immortals, because they were eternal, ever-present in the north sky. The Little Dipper's Kochab and the Big Dipper's Mizar were designated the Oarsmen of Ra—one sitting to either side of Thuban and equidistant from it (see figure 73). That stable portion of the sky, especially the precise northerly space where Thuban sat between Kochab and Mizar, symbolized Eternity in the Egyptian Afterlife.

Learn by Doing

1. **Become a Water Bearer:** How can you better attune yourself with the element of water? Consider this stunning fact: all the clean water that is ever going to be on the Earth is already here. Explore the affirmation-based water-crystal research of Masaru Emoto,[57] or the shamanic work pioneered by Sandra Ingerman that shows that water can be transmuted—changed on a molecular level—through transfiguring ourselves.[58] Engage in spiritual

approaches to water to gain a better working basis for personally creating the shifts in consciousness that you want to see in the world.

2. **Adopt low-flow water practices:** Even renters can easily swap a standard showerhead for a low-flow model. Whether or not your changes affect your water bill, you'll save water for future generations. Locally grown, biodynamic, and vegetarian food choices go hand-in-hand with water-use awareness. There are more food-shopping choices available all the time, and plenty of research to help inform your decisions.

3. **Get involved:** Consider getting involved in larger water issues: join a project that helps to fund a village's drinking water or get active in local water-rights legislation. Most simply, be aware. Aquifer, reservoir, river: where does your water come from *before* it comes from your faucet?

4. **Think like an Aquarian:** Whether the Aquarian Age "officially" starts in four hundred years or tomorrow, that equinoctial shift alone won't provoke peace, beauty, harmony, truth, and love. Invoke your inner Age of Aquarius. How do you envision the future? The future begins in your imagination, your heart, your dreams. It's hard to bring about changes we can't clearly imagine. Dream big; then let your actions reflect those dreams. What we do *today* begins calling forth the changes we want to see.

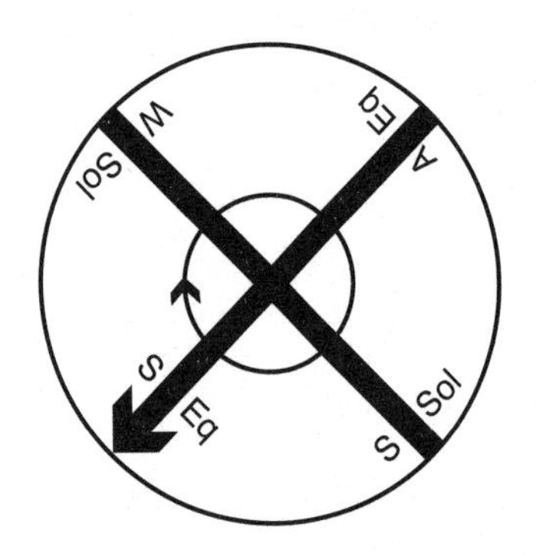

Figure 74. Make an equinox indicator dial.

5. **Make a precession wheel:** Using the modern-boundary version of the zodiac wheel shown in figure 64 or the border-free version shown in figure 67, enlarge the image on a copier, then glue it to sturdy cardboard. Next, copy the circle shown in figure 74. Keep this circle smaller than the first so you can still see the zodiac glyphs. Connect the two through their centers to create an equinox indicator dial. Using the list of dates back on page 80, turn the dial and explore.

6. **Honor the ever-circling Draco:** Here's a tarot reading that friends and I do for each other on birthdays. After shuffling the cards, start with your birthday month and go all the way around, laying down one card for each month (see figure 75). Watch what flows forth as your year progresses. As an alternative (and a reality check), do this reading for the *previous* year, checking the cards that appear in the spread against the events that have already transpired in your life.

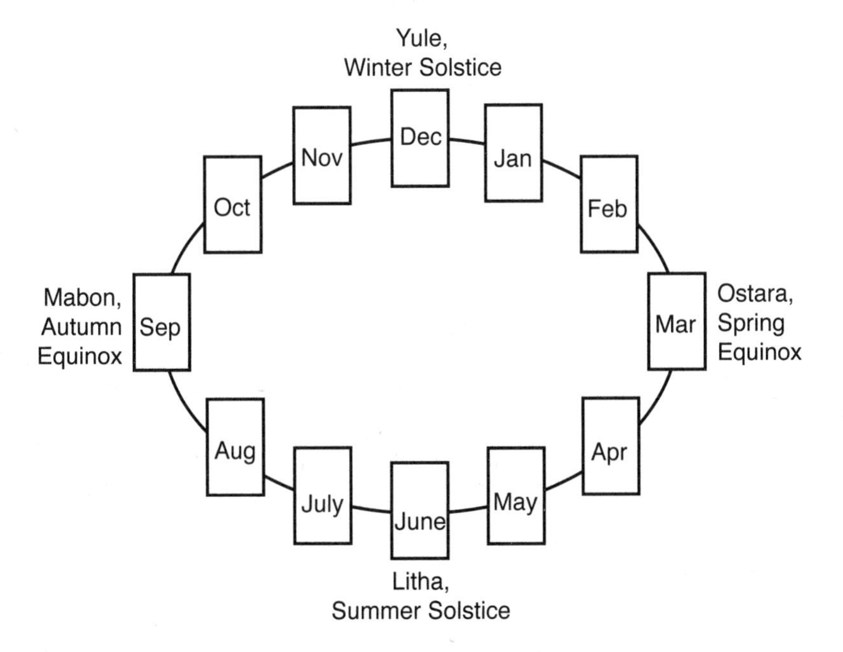

Figure 75. A full-year tarot reading.

Chapter 4

The Moon—Queen of the Night

She with her winding and turning in many and sundry shapes, hath troubled much the wits of the beholders...

—Pliny, *Natural History*, speaking of the moon

In ancient Egypt, a battle raged between the gods. In the twenty-eighth year of Osiris' reign, His jealous brother Set attacked Osiris, sealed Him in a coffin, and tossed it into the Nile, which carried it away. Isis—Orisis' wife and a sister to both brothers—found the coffin, and released and revived Osiris.

But Set wasn't finished. Again He attacked. This time, He dismembered Osiris, scattering the fourteen pieces throughout Egypt. Again, Isis set out to rescue Her husband. She found the lost pieces—everything but Osiris' phallus. Never fear: with Her magic, She crafted Him a new one. Then She reassembled Osiris and again recalled Him to life.

Then Isis took a portion of Osiris' renewed life-essence into Herself: She coupled with Osiris and became pregnant with His child.

Thus Horus, the divine child of Isis and Osiris, was born. He was as wide-seeing as the high-soaring falcon that is His symbol. The Sun is His right eye and the Moon His left. But all too soon, Horus was locked in battle with His uncle,

Set. In their struggle, Set tore out Horus' left eye, ripped it apart, and scattered the small pieces far and wide.

Now Thoth came to Horus' aid. A god of writing and mathematics, a counselor both clever and wise, and a magician in His own right, Thoth found the pieces of Horus' Moon eye and reassembled them. Then, since Thoth also understood the healing arts, He healed—made whole—the left Eye of Horus. That eye became known as the *oudjat,* a sign of wholeness and healing, and of the Moon itself (see figure 76).

Figure 76. The Eye of Horus, the *oudjat*.

Sunshine is wonderful in its light and warmth, and the beauties of sunrise and sunset can be breathtaking. But moonlight—ah, moonlight is where magic happens.

Humans have special connections with the Sun and the Moon, far beyond any familiarity we may come to have with the other planets. We see the Sun and Moon daily, or nearly so. Both are huge and obvious in the sky. Even the smallest children relate to the Moon, its changing shape—round like a ball, curved like a smile—and the softness of its light. The words we have to name the Moon tend to be sweet in the mouth and soft on the ear: *luna, månen, lua, ooljee, mesíc.*

Moonlight reflecting on water is a mystic pathway, beckoning and mysterious.

Sunlight reflecting off water just makes us squint.

Thanks to their perceived parity in size, the Sun and the Moon are often cast as sister and brother, siblings, spouses, or lovers who constantly meet and depart from each other. Like the flip sides of a coin, we see the Sun by day and the Moon by night. At least that's what we *think* we see. A surprising number of adults have told me: "The Moon's not up in the daytime." But half of the time, it is.

The relationship between women and the Moon is ancient. The menstrual cycle is one obvious bond, often moving with the Moon in its rhythm. In folk and metaphysical terminology, the menstrual cycle is referred to as a woman's "Moon-time." The ocean tides respond to the gravitational pull of the Moon, and the female body's inner ocean responds to the Moon as well. Early in human history—before all the artificial lights that keep our nights from being truly dark—women may have ovulated near the Full Moon and bled near the New Moon. Many women find their cycles return to this rhythm if they begin sleeping in total darkness, broken only by the Moon's own Full Moon brightness.[59]

In pregnancy, too, we find strong Moon associations. A woman's womb fills with a growing infant just as the Moon waxes into fullness, then "empties out"—of light, of birthed child. There's timing, too. More births are said to take place on and near Full Moons, with New Moons being the next most common time.

The "ages" of the Moon are also significant—Maiden, Mother, Crone. In lunar terms, a woman's life is echoed by the lunar phases of Waxing Crescent, Full Moon, and Waning Crescent. While this life-journey metaphor can be applied to men as well, it doesn't resonate emotionally as such an intuitive fit.

For all humans, the Dark Moon—when the Moon is unseen, conjunct the Sun—is metaphorically tied to both death and rebirth. A few days after vanishing in the east at sunrise, the slimmest New Crescent reappears in the western sky at sunset. Conjunction with the Sun makes every planet invisible when its turn comes, but the Moon becomes invisible far more often than any of the other planets because it's closest to the Earth and orbits us so rapidly. Seeing the Moon vanish and then reappear each month made it a metaphor for ideas about reincarnation and resurrection.

The Moon in Motion—The Basics

The Moon's movement is tracked in two different ways that take place simultaneously. Both are based on the Moon's unique position as a satellite of Earth: We circle the Sun, and the Moon circles us. In figure 77, the Moon is shown in its Full position, on the far side of the Earth, away from the Sun. We'll meet inner planets Mercury and Venus shortly, but you can see here that neither of them can get far enough from the Sun to be on the far side of Earth.

The first way we can track the Moon is through its *sidereal* period, which views the Moon in relation to the stars. The Moon moves through the entire zodiac in an average of 27.3 days.

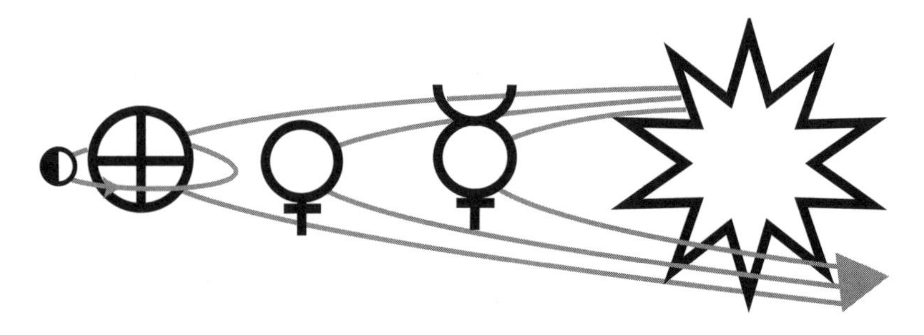

Figure 77. Left to right: Earth with its orbiting Moon, then Venus, Mercury, and the Sun. Hint: We all orbit the Sun the same way that most horse, automobile, and speed-skating races are run in the United States: counterclockwise. The medicine wheel appears again here: It's the traditional astronomical symbol for planet Earth.

The Moon's phases, however (its shifts from New to Full to New), are created via a different timeline—the Moon's *synodic* period, which tracks its place in relation to the Sun. This cycle averages 29.53 days.

Phases are our main way of knowing the Moon. As the Moon orbits the Earth, the Sun's light always illuminates just one side of the Moon, leaving the other half in shadow (see figure 78).

Figure 78 shows how it would look viewed from the side. But the view is different and far more interesting from our earthly vantage point, although only our perspective has changed. In figure 79, the cycle begins with the Dark of the Moon, as Sun and Moon are conjunct, with the Moon so directly in line with the Sun that we can't see it at all. In a day or two, it appears as a New Crescent, first visible in the western sky just after sunset. The solid edge of the Moon's curving shape fits into your curved right hand, like a backward letter C. As the days pass, the Moon is higher each night as the Sun sets, leading to a Full Moon

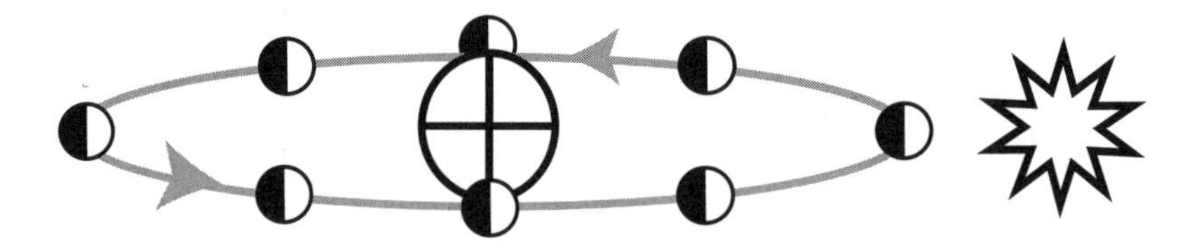

Figure 78. The Sun's light on Earth's orbiting Moon.

rising in the east as the Sun sets in the west, at opposite sides of the sky—that is, as the Sun and Moon are in opposition. The Moon takes two weeks to move from Dark to Full, consistently rising later each day (how *much* later depends on your location and the season).

Figure 80 takes us through the next two-week period. The Full Moon that rose at sunset in Figure 79 is visible all through the night—beautiful!—and sets as the Sun rises. But on each night that follows the Full Moon, the Moon is slimmer, its right-hand side whittled away as it rises after sunset and lingers until dawn. Now the solid edge of the Moon's shape fits your curved left hand, like a letter C. By rising later each day, it catches up with the Sun, shrinks to a tiny crescent, and then finally isn't seen at all. It's conjunct with the Sun again, Dark, directly in line and not visible to us until it reappears in the west at sunset.

We *can* see the Dark Moon—sort of—during a solar eclipse. As the Dark Moon moves directly between the Earth and the Sun, we see the Moon in silhouette for a few minutes, and the brighter stars may appear in the darkened sky. This magical event is possible because, from our perspective, the Sun and Moon are nearly equal in size, 1/2° of space each. A smaller or more distant Moon wouldn't cover our Sun.

Lunar eclipses happen during some Full Moons, when the Earth is so directly placed between the Sun and Moon that its shadow falls across the Moon's face, turning it dark gray or a bloody rust-brown—a possible source of the tradition

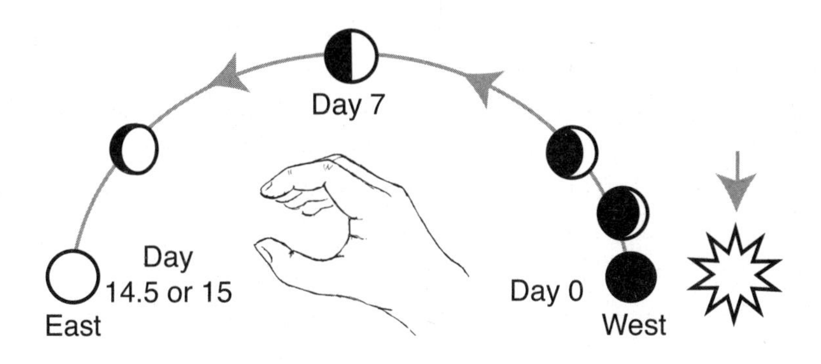

Figure 79. The Waxing Moon. Sunset is on the right, with the first half of the Moon's cycle shown overhead.

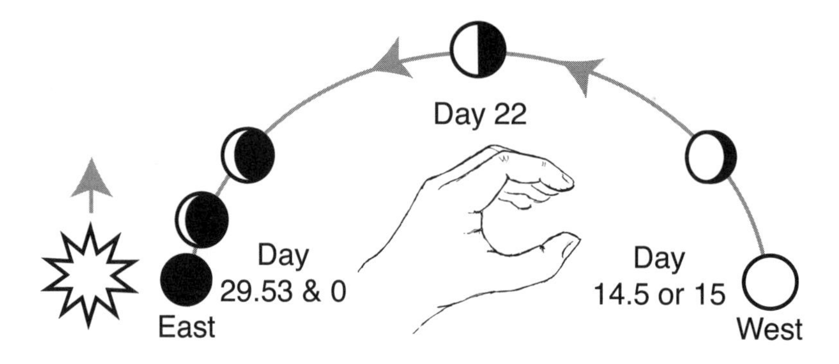

Figure 80. The Waning Moon. Sunrise is on the left, with the second half of the Moon's cycle shown overhead.

that associates the colors white, red, and black with the archetypes of Maiden, Mother, and Crone, with the red symbolizing a woman's fertile years.

Eclipses reveal that which is otherwise invisible. During a solar eclipse, we see the Moon cross directly between the Earth and the Sun, which gradually gave our ancestors an understanding of the Moon's place and orbit. During a lunar eclipse, they saw the Earth's shadow crossing the Moon and came to understand that the Earth is round.

Solar eclipses are only visible to those in the area where the Moon's shadow falls across the Earth. Lunar eclipses, on the other hand, are widely visible. If you can see the Moon, you'll probably see the eclipse (see figure 81).

To put sidereal and synodic cycles together, picture a Full Moon on the Aries-Pisces border. In 27.3 days, the Moon travels through all the other signs and returns to its Aries-Pisces starting point, but it isn't Full again yet. That takes another 2+ days, when our Full Moon is in line with Taurus. The Moon takes an average of 2.25 days to move through each zodiac constellation, so, as a general rule, each Full Moon will be in the next sign. Full Moons occur when the Moon and Sun are opposite each other in the sky, so the Full Moon's place in the zodiac is always opposite that of the Sun. (Sadly, this can't compensate for those pesky irregular-sized constellations.)

The Moon in Motion—The Fine Details

While the Moon is one of the easiest planets to locate in the sky, its motion is one of the trickiest to understand. It's a good thing our ancestors paid attention and helped mark the way.

The Metonic cycle is named for Meton, a 5th-century BCE astronomer from Athens who pinpointed this lunar repetition. Every nineteen years, the Moon returns to the same phase, sign, degree, and declination on the same day of the month, with near exact precision. For example, here are the dates of several Full Moons on or near Winter Solstice (bonus: each of these examples features a lunar eclipse):

> December 21, 1991: Moon nearly to Taurus/Gemini cusp, at 23° 58' N declination
>
> December 21, 2010: Moon nearly to Taurus/Gemini cusp, 23° 32' N
>
> December 21, 2029: Moon on Taurus/Gemini cusp, 22° 24' N
>
> December 20, 2048: Moon on Taurus/Gemini cusp, 22° 11' N[60]

That's calendar-cool, but the Moon's other cycle is far trickier.

Imagine that your tribe does a ritual each month to greet the Full Moon as it rises. One evening, somebody says: "Whoa, wasn't the Full-Faced Goddess on

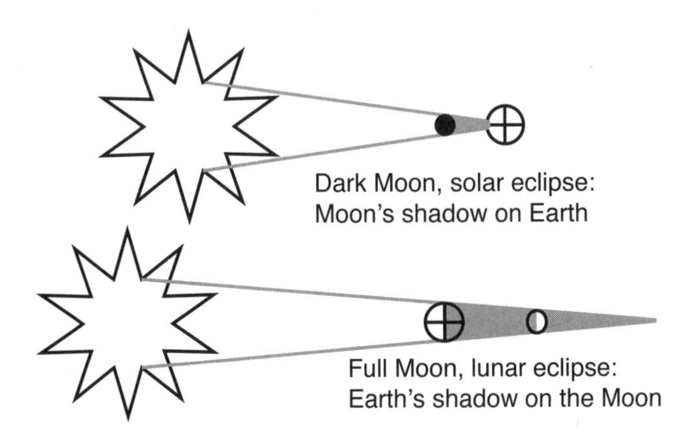

Figure 81. Solar and lunar eclipses.

the *other* side of that standing stone last month?" People then notice that it takes *years* for the Moon to rise on the other side of that standing stone again.

The ancestors were intrigued. They teased out the patterns of the Moon's wandering and then marked it in incredibly creative ways. Given the resulting art and architecture—much of which has only been recognized as lunar-related in recent decades—we can see that honoring the Moon's motion mattered to them a great deal.

While the Sun is steady on its ecliptic path, the Moon meanders. Sometimes her path lies north of the Sun and sometimes south, zigzagging back and forth across the Sun's steady route. This slow-dance pattern of Sun and Moon plays out over an 18.6-year cycle. We see the Sun's extreme north and south points in the solstices. The Moon echoes the Sun's motion and then some, complete with its own solstice-like "standstill"—a multi-month "pause" as it repeatedly hits its extreme positions.

The Sun's Summer Solstice position is 23° 26' N; its Winter Solstice position is 23° 26' S. That's the Sun's *yearly* range. But the Moon swings back and forth north-to-south *each month.* During the maximum period of its 18.6-year cycle, the Moon's position throughout each month ranges from about 29° S to about 29° N. These points are called the Moon's north or south *maximum extremes,* or *major standstills.*[61] For around a year, most months will show these wide swings to either side of the ecliptic, as the rising Moon shifts radically back and forth nearly 60° along the horizon every two weeks. The small Moon symbols used in figure 82 are just that, no more. Here, phases aren't marked, other than Full or New for the eclipses, since Moon phases are independent of this north-to-south extreme cycle.

Midway through that 18.6-year cycle—at about 9.3 years—the Moon reaches its minimum phase, swinging only from about 18° S to 18° N each month (see figure 83). These points are the Moon's *minimum extremes,* also called north and south *minor standstills.*[63] Putting "minimum" together with "extreme" may sound contradictory, but it just means that these are the Moon's most extreme positions in the years when it has the least north-south movement. That's far more bouncing around than the Sun does, but much less Moon variation than in the lunar maximum years.

As with the Sun's angles in figure 19, how extreme these lunar separations look to you will depend on your own north-to-south location.

Figures 82 and 83 both show eclipses that occur when a Full or Dark Moon meets the Sun along the ecliptic. Sun, Moon, and Earth must be exactly in line for a solar eclipse. For a lunar eclipse, the order changes to Sun, Earth, and Moon.

Phases, Extremes, and Solstices

Often, the literature on lunar extremes focuses on solstices and Full Moons, which is misleading. Lunar extremes aren't just Full Moon events, or just solstice events. When the Moon is in an extreme phase of its 18.6-year cycle—at year 9 or year 18—it will reach very near its most extreme point repeatedly throughout that year. For example, figure 82 shows the maximum extremes during March 2006, at 28°41' N and 28°43' S. But it hit those same extreme points again in September 2006 (for two more eclipses) and stayed within 1/2° of those numbers from March 2005 until October 2007. That's thirty-one months. Remember,

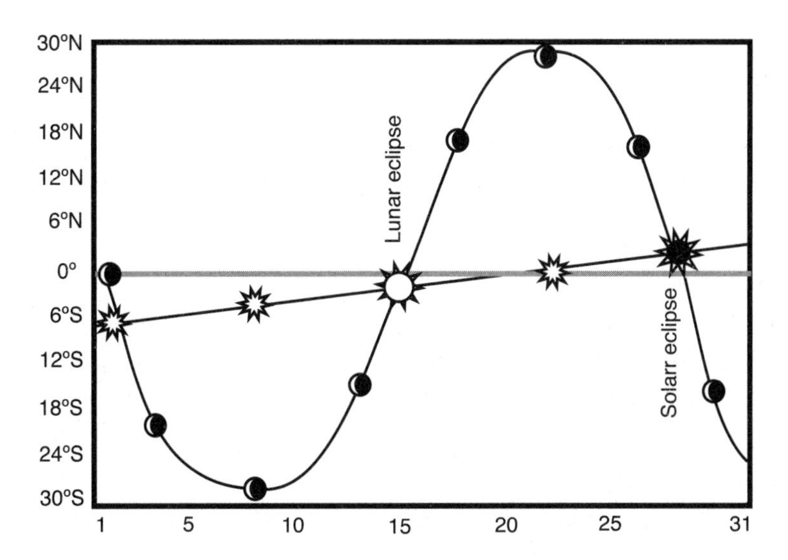

Figure 82. Moon "maximums" over one month, while the Sun moves just a few degrees from south to north. Degrees are shown along the left edge; day of the month is shown along the bottom edge.[62]

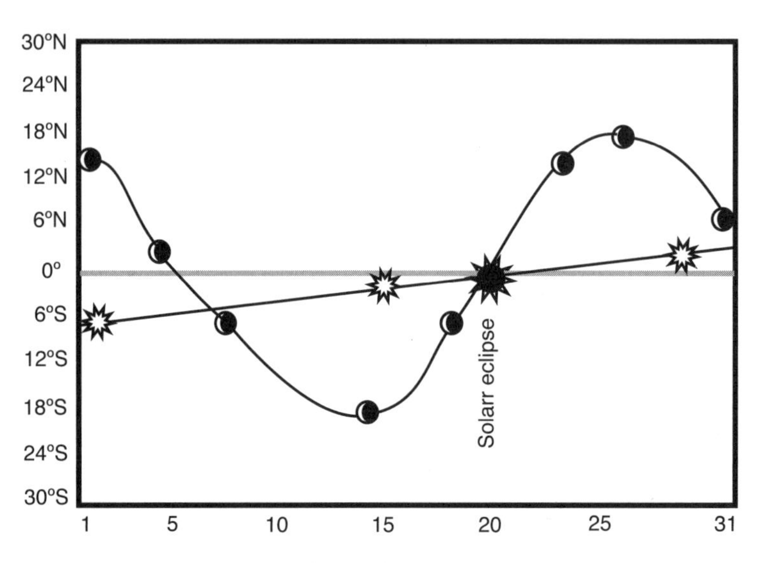

Figure 83. Moon "minimums" during the same month nine years later, while the Sun repeats its same motion south to north.[64]

this is called a lunar *standstill,* as the Moon returns repeatedly to the same extreme positions. This is not a one-night, or even one-month, phenomenon.

If the temple of your hypothetical tribe were aligned to the extremes of the Moon and had any wiggle room at all, you would see several moonrises or moonsets in the key spot during those months, perhaps even some Full Moons. The 18.6-year cycle doesn't affect the Moon's phases, but phases affect visibility. Thus it's nearly impossible to see a New Crescent rising since it's right below the rising Sun, but you may see the same New Crescent set since the Sun's setting has preceded it.

In some years, as your tribe watched the Sun setting in the southwest at the Winter Solstice, they would also see the Full Moon rising in the northeast exactly at its maximum extreme, which was indeed ideal. And if the Moon eclipsed then as well, they might be inspired to enlarge the temple, or to seek more profound ceremonies, or to give precedence to specific deities.

Here are the years for the next few sets of lunar standstills:

Minimum extremes: 2014–2015–2016

Maximum extremes: 2023–2024–2025

Minimum extremes: 2033–2034–2035

Maximum extremes: 2042–2043–2044

Moon Temples and Lunar Architecture

The solstice and equinox points are so steady that observing the skies for even a decade will make clear the regularity of the Sun's pattern.

The Moon, however, is different. With an 18.6-year path as variable as its monthly shapes, how many cycles must be observed before the pattern appears clearly? If the Sun is planetary pop music in 4/4 time, the Moon is planetary bebop jazz. An underlying tempo exists, but can you find it?

A Hopi saying calls the Moon "The Foolish Man Who Runs around with No Home."[65] People have built complex Moon-markers in many locations, as if trying to create some stable mooring for the poor homeless Moon. The Moon's maximum extremes seem to get more attention, probably because these exceed even the Sun's reach from north to south. To mark the Sun's solstice and equinox points, we need six markers: three east-facing positions for sunrises, and another three west-facing for sunsets (see figure 84). To mark all of the Moon's positions—maximum and minimum extremes, north and south, rising and setting—we need to mark eight different positions. Or even ten positions: another two markers are needed for lunar midpoints, rising and setting, unless there were solar equinox markers to share.

Stonehenge, England (51° 11′N, 1° 49′W; c. 3150–1950 BCE)

Stonehenge has the full set of Moon-alignment markers. They are as prominently placed as its solar lines, although, with the 18.6-year cycle, we'll never experience all their views within a single year (see figure 85). This was always true with Moon-marking sites. Extreme sightlines were only functional—accuracy appreciated, cycles duly noted, appropriate rituals observed—gradually, over the long 18.6-year cycle. But people marked them anyway.

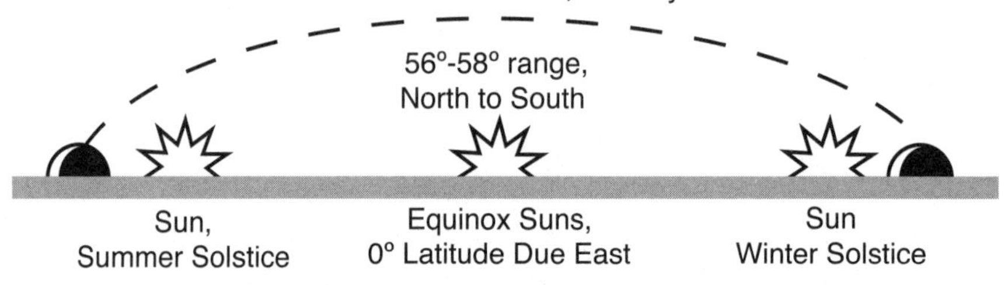

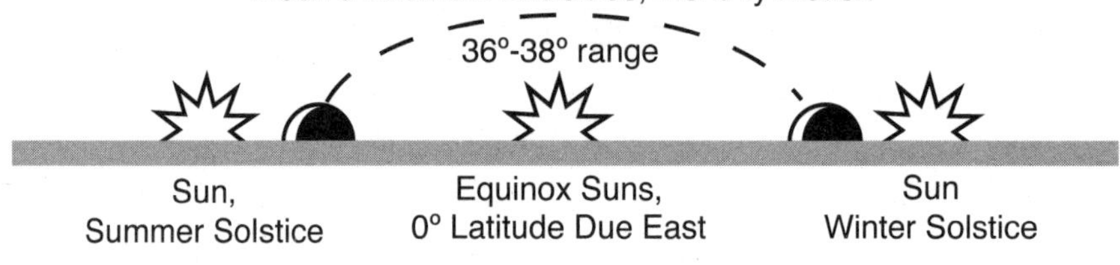

Figure 84. How all this shifting looks along the eastern horizon as the Sun or Moon rises. The Sun's positions represent each year; the Moon's represent year 18 (top) and year 9 (bottom).

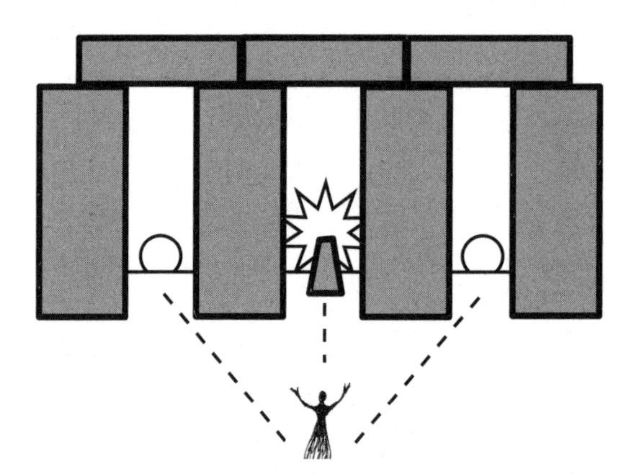

Figure 85. Stonehenge trilithons, facing northeast, with the Summer Solstice sunrise at center (annual), northern maximum extreme moonrise to the left (year 18), and northern minimum extreme moonrise on the right (year 9). The Moon moves *through* the minimum arches regularly, but, in minimum-standstill years, the Moon stops in that archway and then reverses course.

Chaco Canyon, New Mexico, USA (36° 3′ N, 107° 57′ W; c. 850–1140 CE)

The same markers used on Fajada Butte for the solstice/equinox cycle also mark the lunar cycle. Rather than noon sunlight entering between the rock slabs, the rising Moon's light enters from the side, illuminating different portions of the carved spiral (see figure 86).

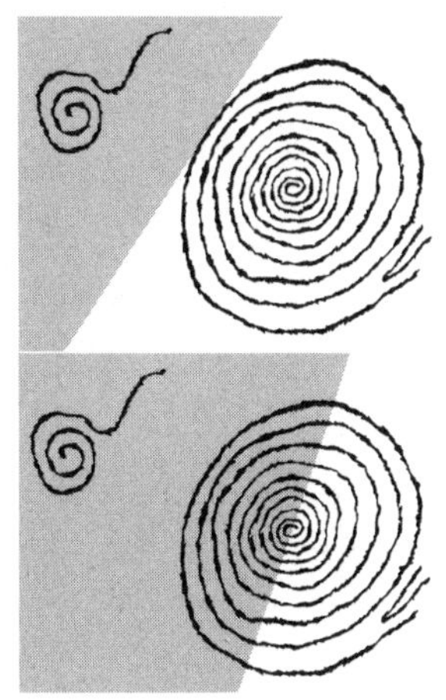

Figure 86. The "Sun dagger" spirals lit by the rising Moon—maximum extreme on top (year 18); minimum extreme below (year 9).

According to Anna Sofaer, long-time Chaco Sun-and-Moon chronicler, this is the "only culture known in the world to align their buildings to the Moon's cycle."[66] Many cultures built observatories with sightlines. The Chacoans constructed entire buildings with either doors and windows or outer walls aligned. And these buildings align with others within the canyon, including sites miles away behind intervening buttes.

Visiting Chaco Canyon now, you often have much of it to yourself—just the buttes, the chamisa, the exquisite stonework, the sky, and lonely little you. But consider that the canyon contains the remains of fifteen Great Kivas that had a capacity of four hundred or more people each and over one hundred smaller kivas with a capacity of fifty to one hundred people each. Even a conservative calculation yields a potential eleven thousand people in kivas, in ceremony.

Eleven thousand. It's a good number to remember if you ever feel lonely in your Moon-watching. In Moon rites, in spirit, I'm often private, but never alone.

Chimney Rock, Southern Colorado, USA (37° 13′ N, 107° 2′ W; c. 850–1125 CE)

Chimney Rock is a Chaco Canyon outlier about ninety miles to the north (145 km). It is culturally associated with, but geographically distinct from, Chaco itself. In about 1050 CE, Chaco immigrants arrived, joining—or taking over?—the established settlement. Advanced-grade sky watchers? They built a Great House on the highest practical land, in the most prominent and defensible position.

From their new digs above the main village, they had an excellent view of Chimney Rock's two rock formations. The stone pillars form a close pair with a

Figure 87. Simulated northern maximum extreme moonrise at Chimney Rock.

gap between them, and the Moon's northern maximum standstill occurs in this natural gateway (see figure 87).

The Great House at Chimney Rock is tree-ring dated to 1076 CE, the year of a major northern standstill. It was expanded in 1093–1094 CE, when the Moon returned to its extreme position.

Octagon, Newark, Ohio, USA (40° 3′ N, 82° 27′ W) and High Banks, near Chillicoathe, Ohio, USA (39° 20′ N, 82° 59′ W)

These two sites were once linked by the Great Hopewell Road (c. 200–500 CE). They are both huge, vast, flat shapes enclosed by low earthen walls. The larger portion of the Octagon site (the upper right portion of figure 88) covers fifty acres (0.20 sq. km). To us non-farmers, that's equivalent to about thirty-eight American football fields laid side to side within the low walls of the Octagon's earthen shape. The idea of this vast space filled with people is mind-boggling, like images of rallies at the Lincoln Memorial. Octagon and High Banks align to moonrises at the northern and southern maximum extremes respectively, events that happen about two weeks apart (see figure 82) at radically different points on the horizon (see figure 84, top). Arial photos from the 1930s show traces of the badly eroded road running about fifty-five miles (34 km) between the two sites. Surveyors in the mid-1800s noticed the road, some 200 feet wide, but didn't grasp its extent.[67] Was it used for ceremonial processions between the two sites?

Ohio alone once held an estimated ten thousand mound sites; about one thousand remain. In and beyond Ohio, America is as rich in mound-builder sites as Britain is in megaliths and stone circles, although the earthworks receive little attention.[68]

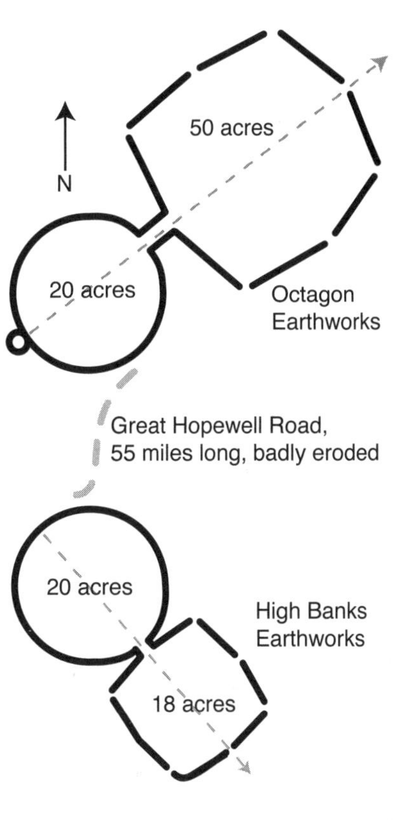

Figure 88. Octagon earthworks (top) and High Bank earthworks (bottom).

Rösaring, Sweden (roughly 59° 30′ N, 17° 30′ E; c. 1700 BCE–1200 CE)

Here, according to recent research, the southern maximum extreme Moon is visible from along the site's ceremonial corridor—its orb resting right atop the mound at the southern end.[69] If Rösaring were 1° farther north, the southern maximum extreme Moon wouldn't crest the horizon at all.

The Sacred Number Eleven, the Eye of Horus, and the Moon's Cycle

I've heard for years that 11 is a sacred number, a perfect number, but the reasons why and the origins of this idea seemed elusive. One possible source relates to the lunar and solar years. The solar year is 365.25 days long. The twelve-cycle lunar year is 354.36 days long. The difference between them is eleven days.

Here's another lunar "eleven," which I'll demonstrate with the Full Moon mentioned previously for December 21, 2010. Add one year and subtract eleven days: December 10, 2011, and another Full Moon. To find last year's Full Moon in any month, start with the current Full Moon, then subtract one year and add eleven days.

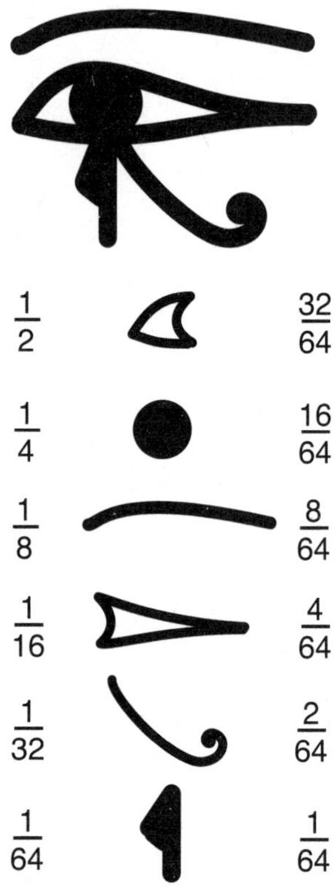

Number tricks bring us back to Isis and Osiris, and the oudjat—the left Moon eye of Horus. It was completely healed, but it was not quite complete.

Consider Osiris' twenty-eight-year reign as a lunar cycle, symbolizing twenty-eight days.[70] The body of Osiris—torn into fourteen pieces and then reassembled—is also lunar, since that's the number of days needed to go from Full Moon to Last Crescent ("torn apart") and New Crescent to Full Moon ("reassembled").

1/2 32/64
1/4 16/64
1/8 8/64
1/16 4/64
1/32 2/64
1/64 1/64

Figure 89. The Eye of Horus (oudjat), divided into fractions.

The Egyptians used the individual portions of the oudjat to symbolize fractions, as shown in figure 89. The curious thing about these six pieces is that you can add them together (1/2 + 1/4 + 1/8 + 1/16 + 1/32 + 1/64) and only get 63/64ths. There's 1/64 lacking. The mythic explanation is that the lone 1/64 is a symbolic honorarium each scholar or mathematician pays to Thoth for His blessing. That's a good story and a good spiritual practice, but here's another explanation that shows why Horus' left eye is so specifically lunar.

The Egyptian standard year was 360 days long (not counting the five extra days that were added so goddess Nut could birth her divine children). That standard 360-day year represents 64/64s.

Divide 360 days by 64 and you get 5.625 days. Then multiply 5.625 days by 63, to represent 63/64ths of 360. The total is 354.375 days, nearly the exact length of twelve lunar cycles (354.36 days). Horus' left eye, the Moon eye that was scattered and magically healed, mathematically represents the days of a lunar year.

Learn by Doing

1. **Connect with the Moon:** Create your own ways to greet and honor the Moon, as a means of building connection with the nighttime sky and of opening your rituals to spontaneity. Here are the bare-bones basics. Whether or not you use these ideas, you can't go wrong if you work from the heart:

   Go outside.
   Breathe deeply and center yourself.
   Pour a libation and scatter bits of bread.
   Speak to the Moon, thanking Her for your life and what's good in it.
   Mean it.
   Enjoy Her light.
   Go inside.

2. **Gather Moon water:** At the next Full Moon, inside or outside, place a bowl of water where it will catch the Moon's light. Remove it the next morning before sunlight strikes it. This is Moon water! Use it for cleansing, drinking, bathing, soaking, or in a spray bottle to add some moist Moon energy to your life.

3. **Go to extremes:** Even when not at an extreme point in its cycle, the Moon swings widely from north to south each month. Find a good vantage point from which to observe this, then notice the range and consider finding a way to mark it.

4. **Moon shadows:** Just as you did with the Sun, you can use your body as a gnomon to cast Moon shadows and mark them on the land. If you were born during the night, time your shadow casting to your birth time. Unlike the Sun shadow for your birthdate and birthtime, your Moon shadow will differ each year over a nineteen-year cycle.

Chapter 5

Mercury—Magical Messenger and Soul Guide

Once upon a time, a young man went down to a crossroads.

Story has it that he met someone there in the dark of night, and that he made a deal. Musical ability? Yes, young man, you shall have it! And how about all the women and whiskey anyone could want? Yes, sir, you can have that, too!

And in the days that followed that dark night, people began hearing something new, as if their ears were on fire. Soon they were talking about the young musician who suddenly had an exceptional talent. Love and drama and lust and hunger oozed from every note he played like honey, ever since that trip to the crossroads.

He came back changed.

In fairy stories, myth, and folklore, one place stands out as the very best location for grand adventures to begin: the crossroads. Whether a formal intersection or a simple fork in the road, when the adventuresome main character of a tale arrives at a crossroads, special things start happening. Driven by a curious spirit, our protagonist leaves the one-dog or one-stoplight town using the only path available; but, as soon as he or she comes to a crossroads—a fork, a "parting of the ways"—active imagination engages. The message? Self, get ready! Choices are coming! It's a new world, with a bounty of information. Communicate, absorb, learn, and choose.

As the little girl said: "We're not in Kansas anymore."

The journey, the quest, the possibilities for transformation—these are the purview of Mercury, the fast-moving planet. Closest to the Sun, Mercury is the hardest planet to spot with the naked eye. Its reputation for being darned near invisible has fueled its mythic associations with cleverness, elusiveness, cunning, intelligence, and inventiveness—all good attributes to have when choices are involved.

In the film *The Wizard of Oz*, Dorothy meets the Scarecrow just as she reaches the yellow brick road's first fork. Her alliance with the Scarecrow soon proves integral to her quest. Although he's already smarter than he realizes, the Scarecrow seeks a brain. In fact, the Scarecrow's quest is very "Mercury," since mental skills, wit, and intelligence fall under the influence of this speedy planet.

In a practical sense, for Dorothy and for us, the choice of route is goal-driven: "Where am I going?" Our personal story's next adventure depends on our answer to this question. But in a larger sense, it's what happens along the path we choose that matters—the experiences, people, and challenges we encounter. Our journeys have the power to change us. At heart, we know this. "Taking the high road" has little to do with elevation and everything to do with tales of honorable behavior fueled by quick-witted awareness and hard-won knowledge.

The "high road" doesn't get much attention in the rich musical genre of "the blues," although cautionary tales abound. Seminal blues musician Robert Johnson (1911–1938) is said to have sold his soul to the Devil in exchange for exceptional guitar-playing and song-writing skills. Some say that Johnson was just okay, until—Shazaam! Suddenly he was brilliant, hence the tale of an unearthly Devil-at-the-crossroads deal. In old tales, "the crossroads" meant outside of town, neither here nor there—good metaphors for transition, for being "between the worlds."

The best times to meet Old Nick were at midnight—the darkest hour, the dividing line between two days—or at twilight—the dividing line between day and night, night and day. Where, exactly? Earthly details aren't important. Supernatural dealings are outside the normal bounds of time and space, so any crossroads can be the "X" that marks the spot.

Musicians, be ready. When the Devil arrives, he'll take your guitar, tune it properly, and play for a while (brilliantly, no doubt) before handing it back. By playing the returned guitar, you take on supernatural gifts, as well as the Faustian cosmic baggage that comes along with them. You get talent with all the trimmings now, but the Devil eventually gets your soul.

Songs like "Me and the Devil Blues," "Cross Road Blues," and "Hellhound On My Trail" fueled the legend of Johnson's deal with the Devil, and his music remains a powerful influence for young blues musicians. He died a violent death at age twenty-seven, poisoned (or shot or stabbed) by a lover's jealous husband. He is honored with gravestones at three different cemeteries—the trickster follows Johnson even in death.

But maybe that crossroads deal-maker wasn't the Devil. There are other powers and other tricksters. In the African pantheon and its related New World traditions, the Trickster is known as Elegua Eshu, Elegba, or Papa Legba. Elegba is an Orisha, one of many spirit deities who each encompass a portion of the divine. Each teaches us life lessons in his or her own particular or quirky way.[71] Elegua, for instance, teaches by playing tricks.

By whatever name, the Trickster is a shape-shifter. Papa Legba can appear as a child, but more often comes as an old man in a broad-brimmed hat, with a crutch or cane (see figure 90). He is the guardian of the crossroads, gatekeeper between the worlds, and master of roads and pathways. One foot rests in the Other World, causing Him to limp, but He's swift of tongue.

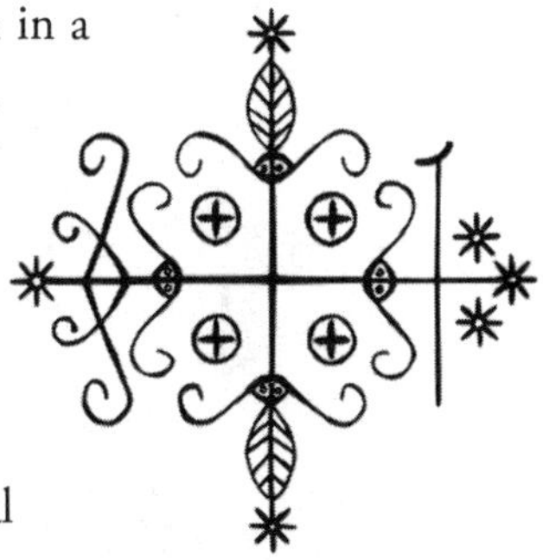

Figure 90. Legba's *veve* symbol, complete with a crutch (seen to the right).

Elegua Eshu's realm is the spoken word. In addition to granting eloquence, He grants comprehension—our crucial ability to understand, not only what's spoken, but what is *meant*. Even in ceremonies dedicated to other Orishas, Elegua Eshu is always the first greeted, invoked, and petitioned, and the final one thanked and bid farewell. He needs to arrive first, because no communications can occur without His aid and blessings to open the way. In this pantheon, Elegua Eshu is associated with the planet Mercury.

Variations on Mercury's Theme

The gods associated with Mercury are all Tricksters characterized by contrasts, contradictions, and unpredictability. Mercury is the god of commerce and successful transactions, but, conversely, He's also the Prince of Thieves. In the New World, He's Coyote. As Shakespeare's clever Mercutio, He is all too human. As Hermes Trismegistus, Thrice-Great holder of esoteric knowledge, He's semi-divine.

Figure 91. The scribe god Thoth with ibis head and writing tablet.

Among the Egyptian gods, He's Thoth, a soul-conducting psychopomp, and the inventor of language and numbers. Brilliant Thoth is often portrayed as a scribe in a human body with an ibis head (see figure 91). The ibis is a sleek waterfowl that has a long, curving beak reminiscent of a pen, especially when the short-legged bird grazes in marshy pastures, bent over the ground like a nearsighted scholar hunched over a scroll.

To the Greeks, Mercury is wing-footed Hermes, child of Zeus, and messenger and herald of the gods (see figure 92). The patron deity of travelers, Hermes also assists in life-and-death transitions as a psychopomp—a conductor of souls. He guides those leaving this life and crossing over to the Other Side, and escorts returning souls back to their new earthly bodies—not as a birth deity (that's a goddess's work), but as the guide of the soul.

Figure 92. Hermes, psychopomp, and messenger and herald of the gods.

In Norse tradition, Odin acted as psychopomp for those who died honorably. He also ruled over war, magic, poetry, quick-wittedness, gambling, and nighttime weather. Odin bartered seriously, trading an eye for a drink from the Well of Wisdom, and suspending Himself in limbo for nine days in exchange for knowledge of the runes. His raven companions Munin ("memory") and Hugin ("thought") helped Him receive and send messages. Sometimes Odin shapeshifted into a raven. Recognize Odin by his spear and wide-brimmed hat; He often has one empty eye socket.

Tokens and Attributes for Mercury and His Deities

Hermes' messenger status was shown by His caduceus, a short staff that served as His badge of office, also used to show that earthly messengers were on official business and should be given right-of-way. This was a sacred responsibility; even your enemy's messenger received safe passage. Early caducei were decorated with fluttering white ribbons, prototype of the white flag of truce. In divine Hermes' case, two snakes are entwined around the stick, as in figure 92. Sometimes wings are added for speed, all echoed in the glyph that represents the planet Mercury (see figure 93).

Figure 93. Mercury's glyph.

The caduceus differs from the rod of Asclepius, which has a single snake spiraling upward (see figure 94). Either can be an expression of kundalini energy, life-force moving along the spine. More pragmatically, snake-entwined staves identify their bearers as divine or divinely gifted.

King Gudea's libation cup shows an early version of the caduceus (see figure 95). The cup honored the king's personal deity, Ningishzida, who was associated with dawn and the horizon.[72] The snakes on the cup are described as copulating, but this entwined motion is a good artistic expression of Mercury's motion.

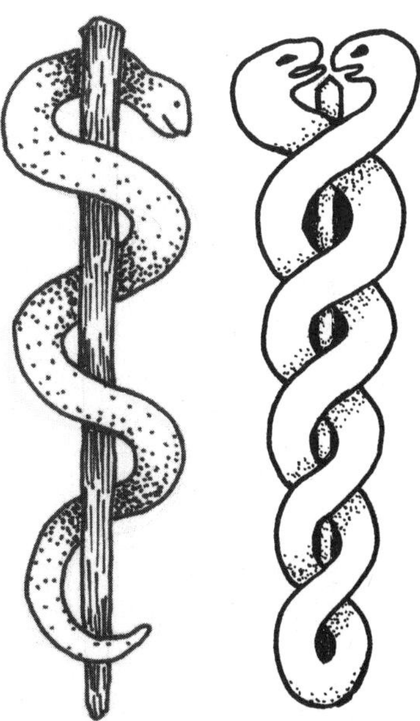

Figure 94. The rod of Asclepius, with a single snake spiraling upward.

Figure 95. Detail of Gudea's libation cup, Sumer, c. 2100 BCE (Louvre, Paris).

Hermes' hat, the *petasos,* was distinctive for its wide, flat brim, and was worn by farmers and travelers to shield them from the Sun. Hermes' hat had wings along the band, which we unconsciously imitate when we tuck feathers into our hatbands. Hermes' hat granted invisibility. Odin's hat is similar and has similar properties, no doubt. Hermes' winged sandals—*talaria*—were golden and indestructible, and granted instantaneous travel.

Black-and-white animals—those with pied coloring—are linked to Mercury. These include the ibis (*Threskiornis aethiopicus*), magpie, woodpecker, skunk, badger, pinto horse, and others. The mismatched colors and patterns in the Jester's garb—called "motley"—reflect Mercury's erratic changeability. Gender-shifting and transgendered deities are another expression of the either-or energy of Mercury. Ninshubar, assistant and messenger to Inanna, is female in some texts and male in others. In the Hindu pantheon, the god-goddess Ardhanarishvara is portrayed as if split down the middle, with mustache and masculine garb on one side, and breast and skirt on the other. The word hermaphrodite blends the names Hermes and Aphrodite, and is used as a biological term for those born with both female and male organs. Odin sought to learn magic from Freyja and was mocked as being "unmanly" as a result. Whenever gender ambiguity is part of a deity's description, a connection to the planet Mercury is likely.

Mercury's "Temples"

Temples were built for this god, but the real focal point for Hermes recognition and worship was the roadside marker called a *herm.* The simplest version of this was a pile of rocks, a loose cairn. As travelers came to a fork in the road, they said their prayers—safe journeys, please!—and added a stone. A more official herm gradually evolved as a blocky rectangle with a head at the top and male genitalia positioned halfway down an otherwise flat side. Sometimes there were two heads atop—male and female—facing opposite directions. These roadside markers also acted as extemporaneous altars for travelers. Want safety, spiritual protection, and guidance? Leave an offering. Locals used the sites to offer prayers for crossroads-type changes in luck, fertility, and fortune. Crossroads were meeting places as well, whether people sought privacy, neutral ground, between-the-worlds ambiance, or simply an equitable location between two villages. The herm marker is the Guardian of the Crossroads.

Mercury's Planetary Motion

All of these associations and attributes—cunning, the snake-entwined staff, the two-faced herm, messengers and psychopomps, young child/old man, speed, crossroads—have their roots in the motion of the planet Mercury. As the closest planet to the Sun, Mercury orbits the fastest, lapping the Sun three times before we Earthlings complete our own year-long loop (see figure 96).

Mercury's inside-track orbit allows two kinds of solar conjunctions. When between the Earth and the Sun, Mercury makes an inferior conjunction (a smaller planet in front of a larger one, marked I in figure 96). Behind the Sun, Mercury makes a superior conjunction (a larger planet in front of a smaller one, marked S in figure 96). Only Mercury and Venus can make both types of conjunction with the Sun, because only Mercury and Venus are inner planets, orbiting the Sun more closely than the Earth itself does.

Imagine a cosmic speed-skating rink with the Sun as the central pillar. We keep changing our seats, moving counterclockwise through the stands; our view directly across the rink changes each time we do. Down on the ice, a skater zips past, moving in the same direction that we are, but going much faster. Approaching from the left (position E in figure 96, to the east of the central pillar, the Sun,

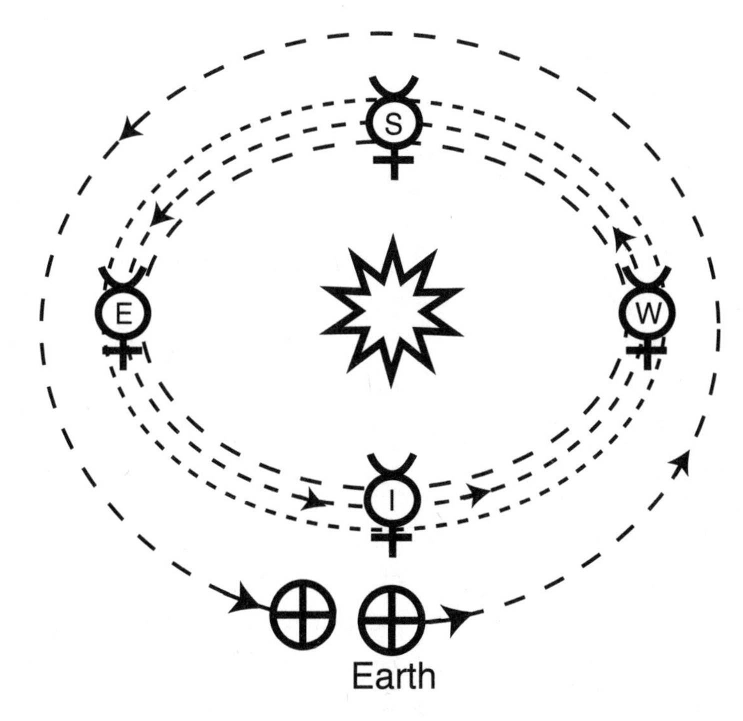

Figure 96. The orbits of Mercury and the Earth, relative to the Sun.

the skater passes directly in front of us (in line with the pillar), then moves away to the right (position W in figure 96, to the west of the pillar). Soon the skater passes directly in front of us on the far side of the rink, in line with the pillar (the Sun) again (position S in figure 96), but now behind it, moving from right to left. These directly-in-line-with-us moments repeat, but since we're changing our seats, the backdrop behind the skater and the central pillar changes as well. As Mercury makes three inferior and three superior conjunctions each year (plus another of either kind if the full cycle straddles the year's end), each conjunction has a different backdrop of zodiacal stars.

Elongation

As with the Dark of the Moon during Moon-Sun conjunctions, we can't see Mercury's conjunctions with the Sun—whether inferior or superior. In both cases, Mercury is in line with the Sun and not visible to the naked eye.

Looking for Mercury high in the night sky doesn't work either. Only outer planets—those with orbits outside of the Earth's orbit—have the option of late-night, overhead viewing. Mercury doesn't, but this planetary Cheshire Cat has other cool moves.

The Moon disappears in the eastern sky and reappears in the west. Mercury vanishes in the east and reappears in the west, then vanishes in the *west* and reappears in the *east.* This quick back-and-forth is part of Mercury's mythology. One Homeric hymn called young Hermes *polytropos,* "of many shifts," a literal description. *Tropos,* you recall, was the origin of *tropic,* for the solstice turning points of the Sun. Mercury has many different, non-solstice turning points, without the Sun's seasonal regularity. This is planetary Whack-A-Mole, and the Trickster planet often eludes us.

To see Mercury, you need to look when it's farthest away from the Sun. This is called the planet's elongation, and Mercury will be either a Morning Star or an Evening Star (see figure 97).

These Morning and Evening Star designations are for planets seen in the east at sunrise or the west as sunset, bright and visible even when the stars aren't. This applies most often to Mercury and Venus, since, as inner planets, these two appear so consistently in dusk and pre-dawn roles.

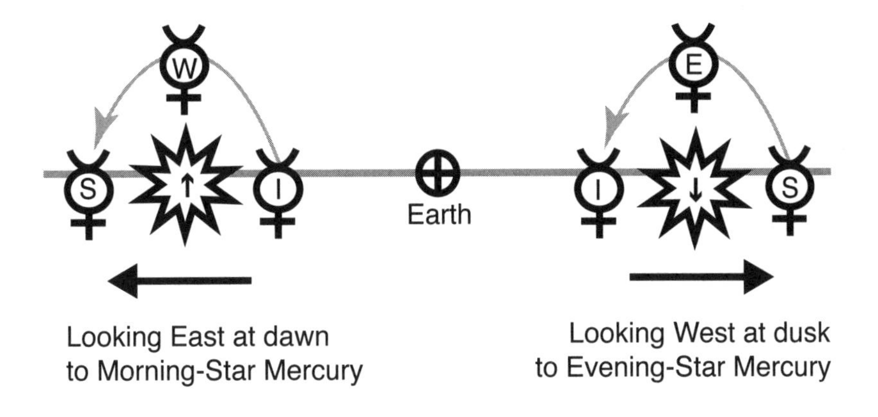

Figure 97. Mercury's dawn and dusk elongations.

Sometimes Mercury is the bright Morning Star, rising shortly before the dawn. This is Mercury as the herald of the new day, new beginnings, and fresh chances for a change of luck.

In the west at sundown as the Evening Star, Mercury is the twilight mediator between day and night, ushering full darkness into the sky while following the Sun down over the horizon. This is Mercury as psychopomp, the travel guide to the Other Side: "Go toward the light!"

In figure 96, our speed-skating rink, we saw Mercury's alignments with the Sun/pillar (the I and S conjunctions). In figure 97, we look at side-to-side elongations (E and W in the figure). Where Mercury is marked W, the planet is west of the Sun, rising before dawn as a Morning Star. Where Mercury is marked E, the planet is east of the Sun, following it, then lingering briefly after sundown as an Evening Star.

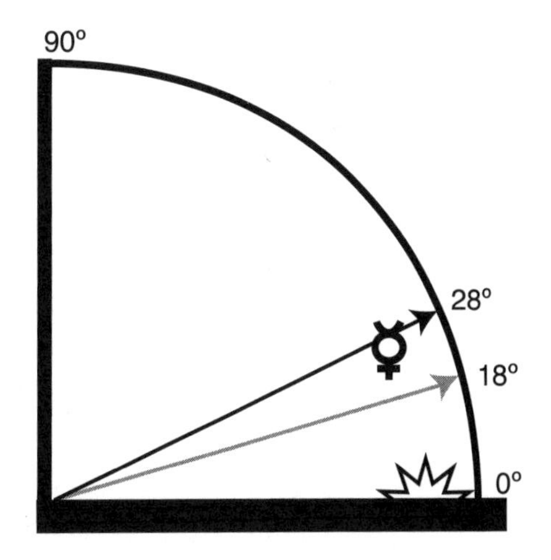

Figure 98. Mercury's maximum separation from the Sun.

Mercury wanders fast, but never far. The greatest elongation Mercury makes from the Sun is 28°, roughly a hand span and a half. Imagine if the Moon only got this high in the sky and never sailed overhead. Often, Mercury's elongation only reaches about 18º; at these times, it is quickly lost in the rising Sun's glare or follows the setting Sun (see figure 98). Your best chance of seeing Mercury is forty to ninety minutes before sunrise (or after sunset), a small window of opportunity.[73] Begin your viewing *before* the exact dates given in appendix A.

Mercury's visibility begins or ends in twilight—times of transition, neither light nor dark, between the worlds of day and night. Twilight is the crossroads of time, a place of changeability and Mercury's habitat.

Cheshire Cat, Whack-a-Mole, Trickster.

Small wonder that Mercury is viewed as both a trickster *and* a loyal messenger. Wandering far away isn't an option; being invisible to mortals is nearly guaranteed. Mercury runs ahead of the Sun as a precocious Morning Star child, or lags behind it as the cane-wielding elder Evening Star. Then, young or old, the supernatural being vanishes.

Travelers setting out before dawn, farmers and other merchants leaving in the wee hours for market—they'll see Mercury. Leave an offering at the herm. In the evening, Mercury may "light you home" as sunset fades, or...

Meet me at the crossroads.

Retrograde Motion

To cap off this bag of tricks, Mercury and the other planets do something that the Sun, Moon, and stars can't—they appear to move backward. This is called retrograde motion.

Despite the circles we use to illustrate the cosmos—round horoscope charts, for example—the planets don't orbit the Sun in circles. Instead, each travels in an ellipse, a circle stretched into an oval. Based on Earth's place in its own orbit and how fast we're moving relative to any other planet, each planet sometimes appears to be moving backward through the zodiac. Picture how, when you are at a stoplight and the car next to you rolls forward, you feel as if you're rolling backward.

We gauge a planet's location and speed by the zodiac backdrop behind it, just as we gauge our driving speed by what we see whizzing past. Imagine driving a flat, perfectly straight stretch of road, with another car far ahead. Is it going the same way you are, or coming toward you, or standing still? Until the road curves, "shifting the backdrop," you can't tell. There's much more to retrograde motion than this in terms of orbital speed and path, and gravitational pull from neighboring objects, but this is the nuts and bolts of it. Planets *seem* to move backward sometimes. They don't really; but *because we're moving too*, it looks as if they do.

Figure 99 shows Mercury's retrograde motion. Follow it using these markers in the figure:

1. Morning Star, western elongation.
2. Mercury directly behind the Sun, superior conjunction, 2/S.
3. Evening Star, eastern elongation, and "backward" motion beginning the retrograde.
4. Mercury directly in between Earth and Sun, in inferior conjunction, 4/I.
5. Morning Star again, and another "stand-still" turn.
6. Mercury clearly "goes direct," forming a superior conjunction again, 6/S.
7. Mercury zooms out ahead of the Sun, an Evening Star, ready to do the whole thing all over again.

As shown here, from one Mercury glyph to the other, this process crosses five constellations. Remember that the Sun's moving during this entire episode, like someone walking slowly while a hyperactive little dog runs circles around them.

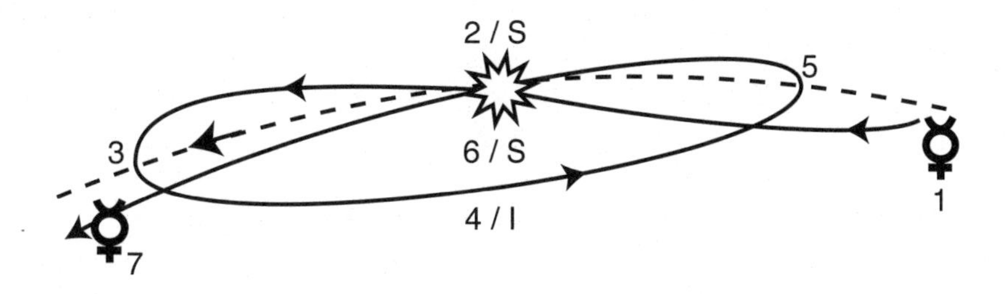

Figure 99. Mercury's retrograde motion.

Learn by Doing

1. **Locate Mercury:** At dawn or dusk, get outside in the twilight and look for Mercury. Use the elongation chart in appendix A or call your local planetarium. Many open-space areas host star-watching nights throughout the year, giving you another potential resource.
2. **Build a herm:** If you're a walker or hiker, take up the practice of building rock herms. Place a stone off to the side at trail crossings, and give thanks to Hermes for keeping you safe so far. Think you've been struggling and suffering? Still, you got here, didn't you? First say thank you, and then ask for continued safe travels. Leave a trail-mix offering or water a nearby plant. You'll notice that many trail crossings already have herms.
3. **Don't blame Mercury:** Can't get through on a phone call? Your computer crashes? A waiter brings the wrong entrée? "Mercury must be retrograde!" Any instance of mental confusion or boggled communication gets blamed on little Mercury. How did this idea become so ingrained in the popular psyche over the past few decades? People who have never seen Mercury and know nothing about its orbit or lore sling blame at it anyway. It seems fair at least to *meet* the planet you're blaming. Challenge: before you blame Mercury again, find it in the sky and make its acquaintance.
4. **Turn crossroads into opportunities:** Forks in the metaphorical road—crossroads—can appear in the form of serious illness, accident, serendipitous meetings, loss, surprising windfalls, or "coincidences." These "road closures" force us to alter our route, but also carry opportunities. The *logos* response of the mind is only one option at these junctures. Often, we have a far more spontaneous or visceral response, and we can take off in a new direction that defies logical explanation. This is mercurial energy in action, and an excellent focus for exploration while Mercury is retrograde.
5. **Heed the message:** The messages we most need to convey and receive may be internal—personal communications with our higher selves or younger selves, with our own sense of the divine, with soul messages percolating within us. Draw on the themes of elusiveness, communication, theft, and

commerce. What is unseen? What needs saying? Or hearing? What has been stolen? What has been negotiated? What's at the heart of the journey? What life crossroads are you facing? Explore these messages using journaling, shamanic journeying, forms of moving meditation and ecstatic dance, dreams, and divination.

6. **Make a Mercury token:** Twilight, light-dark, changeability. If you're courting the Mercury attributes of communicative skill or acumen in commerce, craft yourself a black-and-white or motley Mercury token.

7. **Know your inner gender:** Women seeking to know their Inner Male, men seeking to know their Inner Female, anyone exploring issues of gender—consider making Mercury your ally.

8. **Consider death as transition:** The role of psychopomp—conductor of souls—is gradually reappearing, due in part to hospices, palliative care, and other positive shifts in our attitudes and practices around the end-of-life transition into death. This is a vital change, as important as midwifery and childbirth choices are at the beginning of life. We're only here for a while. No matter what your beliefs about where you'll go *when* you depart, the *act* of departing gives many of us pause. Contemplate what this change means to you.

   The Chalice of Repose Project and shamanic psychopomp work are two clear contemporary reappearances of literal "conductors of souls." The former uses music thanatology designed to support a blessed, peaceful, and conscious passing through prescriptive music via the polyphonic tones of the harp. The Chalice of Repose has its roots in monastic infirmary traditions, which combine medicine, spirituality, and music to support the recipient's transition into death.[74] Shamanic practitioners acting as psychopomps may teach the basics of the journey trance and help the recipient find an entryway to the next world. This allows those in transition to more easily release their hold on physical form when the

Figure 100. The Magician from the Rider-Waite-Smith deck, holding the herald's staff.

time arrives. Shamanic practitioners can also assist deceased souls that are "stuck" in this realm to step across to the Other Side.[75] These and other practitioners are trained and experienced. If you feel called to such work, consider formal training.

9. **Swap the Fool for the Magician:** Many newer tarot decks dress their Fools in a jester's motley, presenting the Fool in fairly playful Mercurial terms. By contrast, Fools in earlier decks were bereft in every way—no sense, no resources, no direction. They were shiftless vagabonds in a much more pejorative sense. And the dog wasn't the Fool's companion. It was biting his leg or buttocks after some upright citizen yelled: "Sic 'em, Canis!" So let's rethink the Fool-as-Jester. The more accurate Jester role falls to the Magician, once known as the Juggler—Le Bateleur, the sleight-of-hand performer, the conjurer. He often holds a short stick, his herald's staff. The Magician is the clever Trickster more appropriate to Mercury. In keeping with his Hermes-like role, the Magician in figure 100 wears a snake as a belt.

Chapter 6

Venus—A Walk with Love, Death, and Rebirth

Inanna placed the shugurra, *the crown of the steppe, on her head.*
She went to the sheepfold, to the shepherd.
She leaned back against the apple tree.
When she leaned against the apple tree, her vulva was wondrous to behold.
Rejoicing at her wondrous vulva, the young woman Inanna applauded herself.[76]

So begins an ancient song now called "Inanna and the God of Wisdom." Inanna, a Great Goddess among the Sumerians, is planning Her visit to Enki, the god of wisdom. When a male hero prepares for a mission, he flexes his muscles, girds himself with weapons and armor, and packs the equipment he'll need for the kind of challenges he anticipates. Inanna admires Her own body in all its luscious sensual detail and applauds Herself. She's ready.

Welcome to the realm of Venus, the planet of contrasting themes: love and war, harmony and fertility, birth and death and rebirth. In ancient Sumer, the planet Venus was personified by Inanna, She of the wondrous vulva. The Sumerians used the same word for "sheepfold," "womb," "vulva," "loins," and "lap," which makes for wonderfully layered imagery.

Venus and Aphrodite

Many of our general ideas about the planet Venus come from Greek mythology and the goddess Aphrodite. The word *aphros* means "sea foam"—from divine sea foam, Aphrodite was born and received her name. Riding on a seashell, she landed on the island of Cyprus, whose name means "copper." This red-gold ore became Aphrodite's metal and, just as weather-aged copper turns green with verdigris, green is Her token color. For the Greeks, She ruled birth and death, time and fate, love and life, all threaded together via sensual mysticism.

Aphrodite was born from the waves already a young woman, beautiful and desirable. Her greatest shrine was in Aphrodisias, where She was worshipped up until the 12th century as the patroness of culture, crafts, and all the arts.

One means of worshipping Aphrodite was as, or through, a *hierodule.* The word means "sacred servant" but tends to be translated as "temple prostitute," missing the larger idea that sexuality and sensuality have spiritual power and energy, as we saw in the story of Shamhat and Enkidu. Joy, beauty, and art have spiritual power and energy as well, so they also come under Aphrodite's purview, although they receive short shrift in many cultures.

The Roman equivalent of Aphrodite is Venus, whose hierodule-priestesses were called *venerii.* That word shares its roots with others, like *veneration,* so let's imagine these women as being very well venerated indeed.

Freyja—Love Goddess of the North

Freyja—which means "lady"—was one of the older northern deities, one of the Vanir. She ruled over all plants, the trees and animals of the forest, love and female sexuality, and magic, and was a recipient of the dead. Lynxes, cats, or bears drew Her chariot. She flew wearing the feathered skin of a falcon, or rode upon Her boar, Hildesvini—"Battle Swine"—who may have been a disguised lover She was keeping handy. Falcon wings, disguised lovers: shape-shifting again.

Freyja's metal is gold; in fact, "Freyja's tears" was a kenning, or poetic metaphor, for gold or amber. She's particularly identified with the necklace Brisingamen, crafted by four metal-master dwarves. She bartered for the necklace, each dwarf receiving a night of Freyja's lovemaking.

Sexual generosity apparently wasn't out of character for Freyja. In the *Lokasenna* ("Loki's Quarrel") portion of the *Poetic Edda*, during a feast, troublesome Loki accuses Freyja of having lain with all of the gods and elves. The response is interesting—no scandal, no tsk-tsking about Freyja, just comments that Loki is being boorish. Her own father, Njord, says: "That's harmless, if, besides a husband, a woman has a lover or someone else."[77] The implication is that women—or at least goddesses—could bestow their favors as they chose.

Next, Loki accuses Freyja of being a sorceress. No one even bothers to yawn. Magic was Her most notable attribute (Freyja taught magic to Odin) and *seidth* work, the trance work akin to shamanic journeying, came under Her purview as well. Freyja's magic covered everything from plants and animals to sexuality, and on to spellwork and prophecy. Like the volvä, Hers is the shadowy skill of spirit-world travel and trance states. Mercury has the role of psychopomp, the conductor of souls who are crossing over, coming and going. By contrast, Freyja is the recipient, the greeter, rather than the conductor of souls. She welcomes new baby souls at birth, and receives those who leave this life.

This brings us back to the planet Venus. In his epic work, *Teutonic Mythology*, Jacob Grimm says: "Whether the planets were named after the great gods, we cannot tell: there is no trace of it to be found even in the North." Grimm then mentions a handful of exceptions, all of which relate to Venus:

> The evening and morning Venus is called *eveningstar*, *morningstar*. . . . There is perhaps more of a mythic meaning in the name *nahtfare* the eveningstar, as the same word is used of the witch or wise-woman out on her midnight jaunt . . .[78]

Here is Freyja as volvä, the seidhr worker we first saw in chapter 0, Her spindle whirling as She moves into trance.

Freyja is associated with the rune *Ehwaz* (see figure 101), meaning "horse," an animal with strong shamanic connotations. Shamans are said to "ride the drum," as if the drumbeats are the sound of pounding hooves, a symbolic horse facilitating the journey into non-ordinary reality. Horse races were held in Freyja's honor at the Winter Solstice.

Figure 101. The rune *Ehwaz*.

In her explication of the runes, *Lady of the Northern Light*, Susan Gitlin-Emmer refers to Ehwaz when she speaks of the imagery of twins in

Scandinavian petroglyphic art—images that mirror each other, or that "appear to be attached at the head."[79] Picture two horses standing nose to nose, or one horse near a wall, nose to nose with its own shadow. This is a good way to portray twins—especially in the metaphysical sense of the self and the shadow self—while simultaneously expressing the idea of travel, or outside-of-self. Gitlin-Emmer notes that, among the Saami people, who have Scandinavia's strongest surviving shamanic culture, the shaman's traveling spirit is called *sueje*, which means "shadow"—an image fitting for inner-planet Venus, which shares Mercury's propensity for twilight appearances.

The Motion of Venus

We'll see more of Venus and her sensual goddesses soon, but to get there, we need a clear grasp on how Venus moves through the sky.

Venus is the third brightest object in the sky. Only the Sun and Moon outshine her. Perhaps that's what originally attracted our ancestors' attention, but the regularity of Venus' motion probably helped as well.

The Moon conjuncts the Sun only by moving between the Sun and the Earth. The outer planets conjunct the Sun only by moving to positions directly behind the Sun. *Only* Mercury and Venus have Sun conjunctions both in front of and behind the Sun. We see Venus make the same loop around the Sun as Mercury, but Venus swings out farther to the left and right. And as with Mercury, we *never* see Venus on the meridian at midnight. Only outer planets can do that.

Just as Diana Prince is never around when Wonder Woman shows up, Venus is either a Morning Star or an Evening Star, never both simultaneously. The Greeks gave Venus a different name for each appearance: Phosphoros, "bringer of light," as a Morning Star; and Hesperus, "the western one," as an Evening Star. The Latin terms are *Lucifer* (which means "light-bringer") and *Vesper*, "evening." Since these names describe where Venus was or what she was doing there, we shouldn't assume that people didn't know they were speaking of a single planet.

Unlike Mercury, Venus is big, bright, and beautiful—easy to spot and, over time, easy to track in its orbit (see figure 102; use the numerical markers to follow the path).

At pre-dawn, looking east, Venus disappears from view for up to fourteen days in her brief inferior conjunction (1). Then, when she's far enough west of the Sun to be visible in the pre-dawn sky, Venus reappears as a brilliant Morning Star (2), with her first post-conjunction heliacal rising. In the following weeks, Venus moves quickly out ahead of the Sun, separating farther to the west (or right) each day, with 47° elongation at most (3). Then, after about nine months as a Morning Star (visible for perhaps eight months of that time), Venus gradually closes the gap with the rising Sun (4), and vanishes when her orbit takes her behind the Sun (5).

Now let's follow Venus' path as an Evening Star (see figure 103; use the numerical markers to follow the route.) Just after sundown, looking west, Venus remains out of sight for up to three months with her superior conjunction (5), but eventually reappears in the West in her heliacal setting, visible just after sunset—a quick blink before she herself sets (6). Gradually, her setting time falls farther behind the Sun's as Venus widens the distance between them, now going far to the east (or left) of the Sun, up to 47° degrees of elongation (7). At this point, Venus is visible for up to three hours after sunset. As an Evening Star, she graces the night sky for roughly eight months, then closes her gap with the Sun

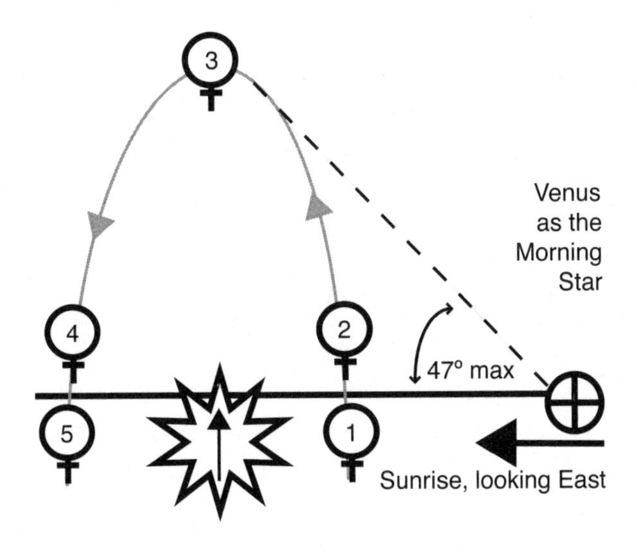

Figure 102. The path of Venus as a Morning Star.

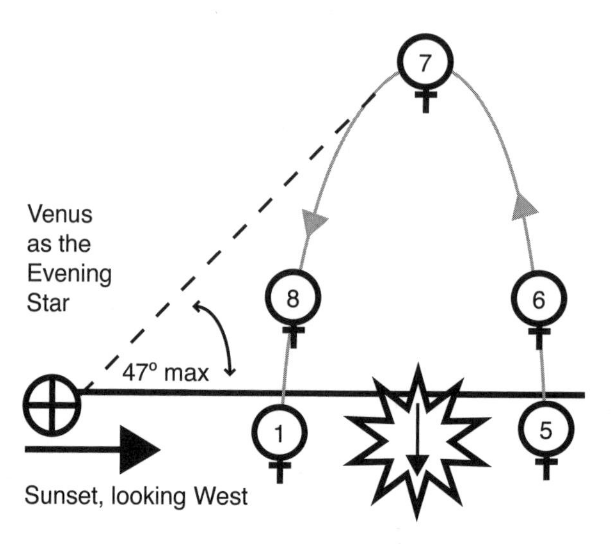

Figure 103. The path of Venus as an Evening Star.

again (8), and vanishes back into an inferior conjunction. She reappears soon in the eastern pre-dawn sky as a Morning Star (see marker 1 in figure 102). This is a portion of Venus' retrograde motion, replicating that of Mercury shown in figure 99.

Due to the relative positions of the Sun, Venus, and the Earth, we see Venus trace a pentacle, a five-pointed star, in the sky through a series of solar conjunctions. Its shaping is nearly perfect; its repeating pattern and schedule are precise. Many early civilizations understood this cycle, tracking Venus' motion with a precision worthy of her.[80]

The Sun and Venus meet in a conjunction, either inferior or superior, every 584 days on average. Figure 104 shows the locations of Venus' inferior conjunctions (expressed astrologically); figure 105 shows her superior conjunctions. If you mark six conjunctions of either type, then connect them in chronological order, you'll see the pentacle. The Venus pentacle cycle covers eight years (minus about three days), and the planet returns to within 2° or 3° of her starting point.

We can't see Venus during her Sun conjunctions, but we can predict the conjunctions by using Venus' maximum elongations as a Morning or Evening Star.

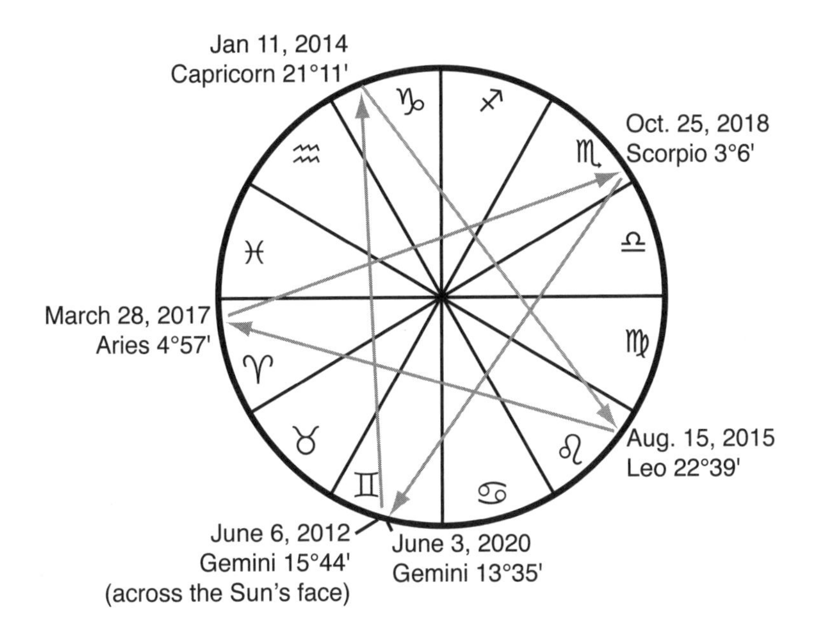

Figure 104. Venus-Sun inferior conjunctions.

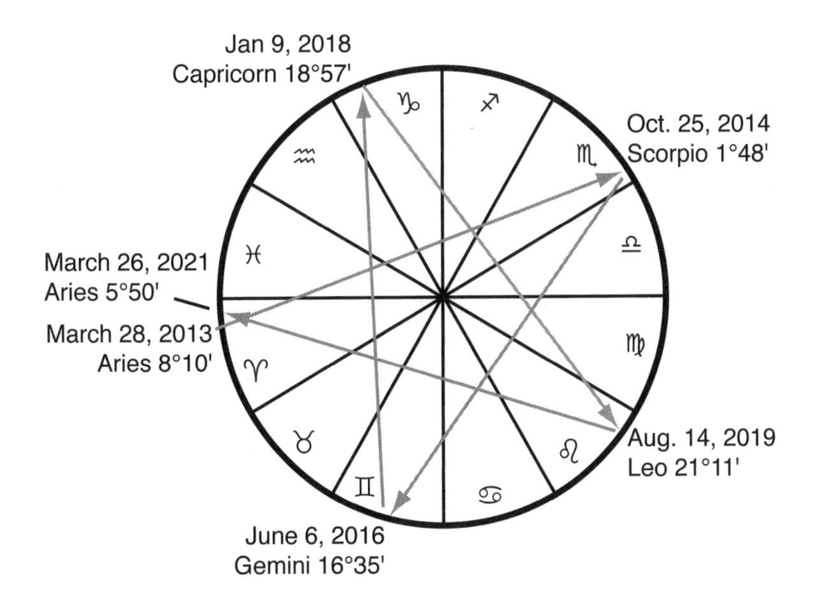

Figure 105. Venus-Sun superior conjunctions.

Here's a partial list of Venus elongations. Notice that the dates in figures 104 and 105 reappear here, either exact or close, but one or three years off. This is because Venus' Evening Star elongations take place one year *before* a superior conjunction, and three years *after* an inferior conjunction, while Venus' Morning Star elongations happen one year *after* a superior conjunction, and three years *before* an inferior conjunction.

Morning star	March 22, 2014	46° 36' west of the Sun (Sup conj +1yr)
Evening star	June 6, 2015	45° 24' east of the Sun (Inf conj +3yr)
Morning star	October 26, 2015	46° 24' west of the Sun (Sup conj +1yr)
Evening star	January 12, 2017	47° 6' east of the Sun (Inf conj +3yr)
Morning star	June 3, 2017	45° 54' west of the Sun (Sup conj +1yr)
Evening star	August 17, 2018	45° 54' east of the Sun (Inf conj +3yr)

I found this correlation from the conjunction side, but the formula works just as well going the other way, of course. If you track Venus' greatest elongations *from* the Sun, you can use those dates to predict Venus' conjunctions *with* the Sun. Stunning, really. (There's a list of Venus elongations from 2010 to 2050 CE in appendix B.)

Despite the five-pointed star in Venus' orbit, Inanna's symbol among the Sumerians was an *eight*-pointed star, or rosette (see figure 106). This isn't contradictory. The rosette may express Venus' eight-year cycle; it may also refer to Venus' roughly eight months of visibility as an Evening or Morning Star. These possibilities aside, the eight-armed rosette is definitely a graphic depiction of Venus' maximum separation from the Sun. Her average elongations range from 45° to 47°; a circle divided into 45° segments yields eight segments, like the eight-spoked rosette. Too esoteric for the average ancient citizen? Probably not. An eighth is half of a half, halved again, which is several steps short of the calculation of sixty-fourths we saw with the Eye of Horus. In the tarot, the Major Arcana's Star card is often depicted as an eight-pointed rosette, and some writers associate the card with Venus (see figure 107).[81]

Figure 106. An eight-pointed rosette with a 45° angle marked.

Tracked through either type of conjunction, Venus keeps right on repeating her pentacle pattern and her stunningly visible elongations. And our ancestors were watching every move.

Morning Star, Evening Star—Where Does the Romance Come In?

The connection of the Evening Star to romance and lovemaking is pervasive. Maybe folks felt more romantic when Venus emerged from her long superior conjunction to light the evening sky as the first "star." Perhaps cultural taboos discouraged sexual relations during her invisible conjunctions.[82] Either way, let's imagine celebratory passion when the Evening Star finally reappeared. This amorous image parallels songs of praise honoring Venus as Inanna. In "The Lady of the Evening"—one of the first known songs of love—Venus, as the Evening Star, is sighted. One exquisitely sensual translation tells us that men and women both tidy themselves, and "My Lady makes them all hurry to their sleeping places. . . . There is great joy in Sumer. . . . The young man makes love to his beloved."[83]

Figure 107. Major Arcana Star card (XVII). Note that the star is represented as an eight-pointed rosette.

In the texts, which date back to at least 2000 BCE, Inanna prepares to meet shepherd-king Dumuzi, "the wild bull," who comes to Her "out of the poplar leaves":

> My vulva, the horn,
> The Boat of Heaven,
> Is full of eagerness like the young moon.
> My untilled land lies fallow...
> Who will plow my vulva?
> Who will station the ox there? ...
> Last night as I, the Queen of Heaven, was shining bright...
> He met me—he met me! ...
> Wild bull, Dumuzi, make your milk sweet and thick...
> Fill my holy churn with honey cheese...

> The shepherd Dumuzi filled my lap with cream and milk…
> He quickened my narrow boat with milk.
> He took his pleasure of me.
> He brought me into his house…
> O Dumuzi! Your fullness is my delight!…[84]

Pleasure, holy churns, and joyful sexual imagery aside, we know the Sumerians equated Inanna with Venus, but does this also describe Venus near a new Crescent Moon?

Reemerging from its own dark phase, the Moon's tiny crescent rides so low in the sunset sky that we may only see it between tree branches—"[he] rose to me out of the poplar leaves"[85]—before it quickly sets. With the bright horns of a wild bull, the Moon waxes slightly wider over the next several days and has more sky time with young Venus each evening before they both set. On each successive night, the Moon's crescent bowl becomes more filled "with milk and cream," moving from newest crescent toward first quarter (see marker 1 in figure 108).

And while the young Moon is the wild bull, it is simultaneously Inanna's lap and Her Boat of Heaven, "eager as the young moon," waiting to be so exuberantly filled. This is the beauty of metaphor and *double entendre:* We don't have to stick with a single image or symbol, but instead can overlap, layer, and braid them. Multiple meanings are employed simultaneously—complex flavors for the poetic taste buds—in a wave of competing but complementary images.

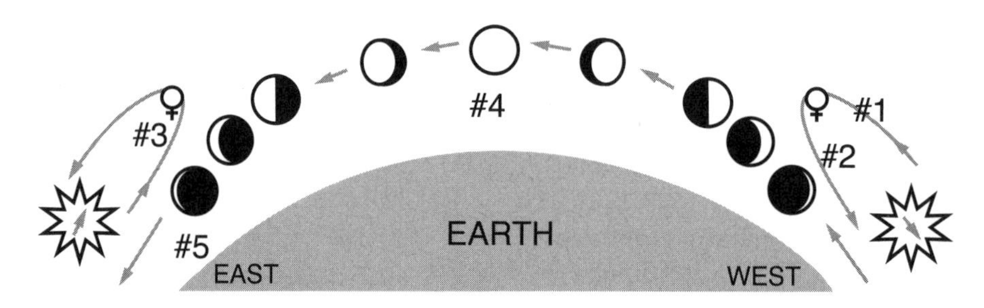

Figure 108. The path of Venus depicting Inanna (Venus) and Dumuzi (the Moon) together in the night sky.

While the Moon is youngest, its lapis-blue roundness is faintly visible via "earthshine," even when only a slim crescent edge is catching sunlight. During that time, the Moon may bring Venus "into his house" by edging close alongside her in the sky, or even occult her by moving directly between her and the Earth, so that Venus is eclipsed by the Moon.

This especially intimate celestial wooing doesn't occur with each Evening Star cycle. It depends on the Moon's north or south position in its own multi-year cycle. Thus it is a special time indeed when these two meet so closely and dramatically.

The Inanna-and-Dumuzi extract above is enough to get a sense of their sexual frisson. With the Sumerians writing erotica like this for two celestial lovers, we'd be foolish to think they weren't mirroring some of this love-play at home, by the light of the young Moon and the Evening Star.

Birth—and Rebirth

Venus' traditional associations aren't simply to love, but to fertility, and to women specifically. Isis, Astarte, Inanna, Ishtar, Aphrodite—these goddesses' sacred connections to women come directly from the planet Venus herself. Venus' life span as either a Morning or Evening Star, added to the length of her inferior conjunction, roughly matches the length of a human pregnancy, which averages 260 to 265 days from conception.[86] Pregnancies begun near Venus' reappearance as an Evening Star come to term around the time of her rebirth as a Morning Star.

The Babylonian ideogram for "mother" was a combination of shapes meaning "deity" and "house," as a mother miraculously offers herself as first "house" to her child.[87] Other titles granted to the Great Goddess, and to Venus in particular, include Lady of the Womb, Lady of Form-Giving, Carpenter of the Insides, Lady Life-Giver, Lady Potter, and Lady of the Embryo.[88] These are all in keeping with Venus as a Morning Star, herald of birth, dawn, and new beginnings.

But Venus plays an even more profound role as a symbol of *re*birth. In the epic known as "The Descent of Inanna," this goddess is the first being of either gender known to undertake an initiatory journey of self-discovery, sacrifice, and resurrection.[89]

The tale begins when Inanna "open[s] her ear to the Great Below,"[90] seeking knowledge of the underworld, the Kingdom of the Dead ruled by Her sister, Ereshkigal. First, Inanna tells Her loyal servant Ninshubar how to mount a rescue if She doesn't return promptly. Then, taking up the symbols of Her rulership, Inanna departs for the underworld (Venus' western disappearance—as an Evening Star—into inferior conjunction; see marker 2 in figure 108). "I am Inanna, Queen of Heaven, on my way to the East," She declares.[91] And Venus *will* next be seen in the east. Interesting.

Figure 109. The Egyptian hieroglyph for the "underworld."

Like the Dark of the Moon, this is the Dark of Venus, the unseen time of conjunction. An Egyptian hieroglyph for "underworld" was a pentacle within a circle (see figure 109),[92] an accurate expression of Venus' pentacle-marking conjunctions, during which she is indeed unseen after disappearing into the west—the direction of farewell in many traditions. And isn't she going underground, into the underworld?

Repeatedly challenged by Ereshkigal's chief gatekeeper, Inanna gives up a token of Her rulership at each of seven gates to the underworld. Her *shugurra* crown goes first, then Her necklace of lapis beads, followed by the double strand of lapis across Her breast, then Her breastplate, then Her gold bracelet, then Her lapis measuring rod and line, and finally Her royal robe. Lapis lazuli—deep blue with flashes of golden pyrite—is appropriate here, because it's like the night sky that Venus-Inanna relinquishes by going "underground." In a larger sense, those things that establish Her identity are left behind, just as, in personal cathartic moments, our concepts of identity and sense of self are shaken.

"Naked and bowed low," Inanna comes into Ereshkigal's presence. The underworld sister is relentless. Those who enter Her domain belong to Ereshkigal alone. The Queen of the Below strikes down Inanna, transforming Her into a corpse—"a piece of rotting meat"—to hang on the wall like a grisly trophy.[93]

As instructed, Ninshubar waits for three days, then raises the alarm. She laments; She drums (as a good shamanic assistant should); She mourns; and finally, She actively seeks help. Enki, God of Wisdom, tricks Ereshkigal into relinquishing the corpse. Inanna is restored to life, and Her seven possessions are returned as She passes back through the seven gates. The goddess Inanna is about to ascend from the underworld (Venus' reappearance as Morning Star; see marker 3 in figure 108) except for one pesky detail. Ereshkigal demands that someone take Inanna's place.

Ereshkigal's servants, the relentless *galla,* will make sure that Ereshkigal gets a replacement corpse. Inanna refuses to let the galla take Ninshubar or anyone else who's been loyal to Her, but then She spies Her lover-husband, Dumuzi:

"Inanna fastened on Dumuzi the eye of death . . ."[94]

Everyone else was mourning Inanna's absence, but Dumuzi had instead been "shining" on His "magnificent throne." Perhaps as the Full Moon, high in the sky (see marker 4 in figure 108), traveling boldly across the sky each night while Venus was Sun-conjunct languishing in the underworld?

Dumuzi hasn't betrayed Inanna by usurping Her place or by neglecting to rescue Her. Remember, that was always Ninshubar's role. Dumuzi's downfall is that He has failed to mourn for Inanna.

> The *galla* seized [Dumuzi] by his thighs.
> They poured milk out of his seven churns . . .
> [Dumuzi's] shepherd's crook has disappeared . . .
> The churn lies silent; no milk is poured.
> The cup lies shattered; Dumuzi is no more.[95]

Seven churns? The Moon's last quarter takes seven days to pour out its "milk" as the Moon's shape is pared away. The Waning Moon vanishes into the dawn, while every morning the triumphant Inanna-Venus rises earlier, brighter and higher in the pre-dawn sky. Venus and the Waning Crescent Moon pass each other in the pre-dawn sky, bound in opposite directions. But Dumuzi is no more (see marker 5 in figure 108). Yes, He'll reappear in three days in the west, but Venus is a Morning Star now, at the other side of the heavens, and in no mood for mush and love songs. When Venus is a Morning Star, the Moon *always* meets Her in a diminished state, waning and contrite.

Versions of this story repeat in other times and places, with different characters—the first of numerous other sacrificial and born-anew goddesses and gods, heroes and heroines, guides and teachers. Did our wisest storytellers overlay

Figure 110. Newgrange entryway.

significant themes onto the planetary gods, whose choreography seemed to mimic our human plights?

This much we do know: When this goddess appeared to the Sumerians, they honored Her. They sang songs of praise, played drums and tambourines, and made love—that most personal rhythmic event. Their astronomers counted days and tracked elongations. They all welcomed and acknowledged Inanna in their own ways, which weren't contradictory—math for tracking Her heavenly motion, and poetic metaphor to convey Her story to an observant population that watched Her manifestation in the sky.

Tracking Venus' appearances, disappearances, and star-shaped pattern are the work of logic and relate to our quest for understanding through information analysis. The *experience* of Venus' motion is the domain of spirit, as we open ourselves to more intuitive ways of knowing. Venus—goddess and planet—fuses these paths.

Newgrange—Venus Marked and Honored (Boyne Valley, Ireland: 53°41' N, 6°28' W; 3100–2900 BCE)

We already saw the Newgrange entrance as the solstice Sun streamed in. But what if the light box wasn't intended to emphasize *sun*light? Alternative theories suggest that Venus was its real focus.[96]

Figure 111. Close-up of the Newgrange light box's decoration.

Every eighth year, when Venus rises as a Morning Star before the Winter Solstice Sun, the light box catches Venus' cool light and briefly illuminates even the deepest portions of the chamber.[97]

Over the entryway, there's a subtle carving on the lintel above the light box, shown in figure 110 with a portion of the great spiral-carved stone positioned in front of the door. In figure 111, I've highlighted the light box's carved panel to show the detail, which is otherwise lost in shadow. Like the Mescalero Apache's quartered circle that expresses a year's cycle, theories hold that each X-block here represents a year, as an encoded reminder to heed Venus' eight-year cycle. This would be in keeping with a folk tradition that speaks of the Morning Star's light entering the chamber.[98]

Newgrange is known to be full of sophisticated and well-engineered marvels. There's a passive drainage system built *within* its five-thousand-year-old roof.[99] Venus alignments may have been child's play to those who built the temple.

Learn by Doing

1. **Admire and applaud yourself:** From childhood to adulthood, we all have many chances to absorb body-image negativity. If you got a dose of that horrible self-loathing, Inanna is an amazing role model for losing it. Lady readers, mirror the self-affirming practice of Inanna. Go find an apple tree and lean back against it. Admire and applaud yourself—head to toe and everything in between. Gentleman readers, personalize and adapt this practice as you will.

2. **Find the daytime Venus:** Venus is visible during the day. At her greatest elongation, you can find her in the pre-dawn sky and watch her into the daylight hours, although she's much less visible then, since her brightness provides little contrast against clouds and pale-blue daytime skies.

3. **See (carefully!) the transit of Venus:** During her inferior conjunction, Venus very occasionally lines up so precisely with the Sun that she actually crosses the Sun's face, left to right, like a tiny speck. This very rare event occurred on June 8, 2004, and will happen again on June 6, 2012. The phenomenon won't repeat itself until December 2117, so be sure to see it in 2012.[100] Check with planetariums in your area for viewing options, and don't look directly at the Sun.

4. **Interpret Venus in the tarot:** The Major Arcana's Star card and its eight-pointed rosette can represent Venus. If this card appears in a reading, consider applying Venus' challenge-and-rebirth metaphors—or Inanna's sexual self-assurance—to your interpretation.

Chapter 7

Mars—A Planetary Rebel?

Imagine if every time your name were mentioned, people started mumbling, "War...belligerence...struggle...challenge...aggression...temper...anger...argument...conflict." There's an apprehensive murmur as tension mounts, everyone waiting to see what kind of trouble you'll start. And you *will* start trouble, right? Like all those pesky movie Martians who come to Earth to suck out our brains and pillage our resources?

We humans have a capacity for violent behavior, and over many generations, that mayhem became symbolically attached to Mars. Violence—enacting it, protesting it, inflicting it, escaping it, and healing from it—permeates our entertainment themes. Images of real turmoil from around the world are broadcast and emailed instantly, as if they were happening to us, and our shared anxiety rises. Fight or flight, fight or flight, fight or flight...

Before we can manifest something, we must first imagine it.

What results can we anticipate when so much of our mental, emotional, and imaginative energies are wrapped up in violent themes?

This is some of the baggage that comes with Mars—drumming in our brains like a negative-emotion mantra. Ouch.

I asked a wise, astrology-savvy friend: "What image do you use for the deity Mars?"

Her reply? "The Green Man."

The goddess Juno was sorrowing, offended that Jupiter had birthed a child—Minerva—who was not born of woman, but instead sprang from the head of Jupiter Himself.

Juno called upon Her handmaiden Flora to bring Her a flower. Flora, the goddess of flowering plants and springtime, did as Juno asked, and Juno clasped the flower passionately to Her breast. Through the intense longing of Her embrace and Her own generative deity, Juno became pregnant. The child She bore was Mars, who inherited His flower-father's fertile green energies—and His mother's anger with Jupiter.

Figure 112. Juno with Her flower and infant Mars, from a Roman coin, c. 222–235 CE.

So here's where we begin—with a family portrait of a flower-father, an angry goddess-mother, and a divine baby (see figure 112).

Mars' role was originally one of fertility and defense. He caused the plants to grow, protected cattle and fields, and established and maintained good boundaries around these settled places. He fathered Romulus and Remus, who in turn became the founders of Rome. The month of March—first month of the Roman year—is named for Mars, and that season's earliest associations were with planting new crops. Mars helped with that—via prayers and offerings—and was called upon when divine defense was needed. Gradually, this changed, and March became the month to muster the troops. Why just defend your own fields when you can go out and conquer someone else's? But this is a later Rome and a later Mars, farther from the wild woods, the fertility of the earth, and the life-force of nature.

The Unique Red Planet

Mars is often called "the red planet" and, in modern photographs taken with space probes and telescopes, the planet indeed is shown to be a rusty orange-red. But if you're looking for something distinctly red using only the naked eye, you'll miss Mars, which just looks warm and peachy-orange in comparison to the other planets and stars around it. But the more stars you can see, the more Mars' warmer tone stands out from the crowd in contrast. This warmth and redness have been spun into Mars' lore and symbolism as blood, iron, temper, fiery nature, and passion.

We can attribute some of the god Mars' storied combativeness and contrariness to the planet Mars' own motion in the skies. A crucial part of Mars' identity stems from planetary order. Mercury and Venus, as we've seen, are never far from the Sun. Next is the Earth. Then we come to Mars and something new is introduced: opposition (see figure 113).

Opposition occurs when the Earth is positioned between two other planets at opposite sides of the sky. When the Sun and Moon are in opposition, there's a Full Moon, but that occurs because the Moon is orbiting *us*. Moving away from the Sun in position and length of orbit, Mars is the first planet out from the Sun capable of being in opposition to it. Hence, Mars the rebel.

Astrologically, we still chart the heavens as if we Earthlings were the center of the universe with the Sun circling us (see figure 114). This is not accurate, but that's how it looks from our perspective.

But Mars has another surprise for us. Although (and because) Mars is moving much faster than either Jupiter or Saturn, it opposes the Sun *less* often than either of these slower planets. In fact, Mars has solar conjunctions and oppositions less frequently than any other planet. Mars comes close to repeating its moves over a fifteen-year cycle, but not *really* close; you won't see the cycle clearly unless you watch it over a mind-numbing seventy-nine years, as shown in appendix C.

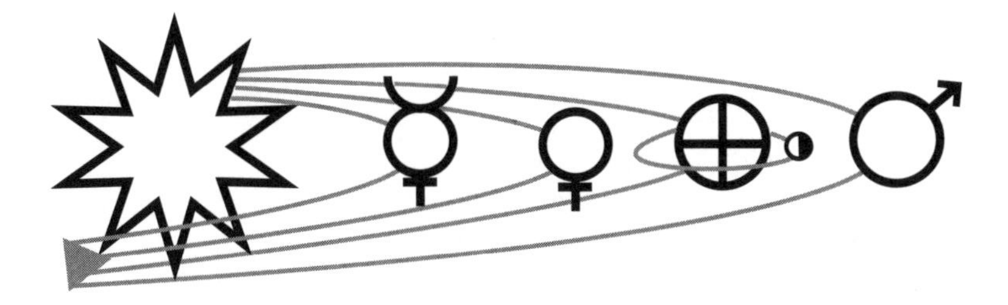

Figure 113. Mars' place in the planetary order.

After Venus' orderly eight-year star dance, it's small wonder that Mars seems so contrary. Even Johannes Kepler, modern master of comprehending planetary motion, struggled to formulate a theory to express the movement of Mars.[101] As we'll see in subsequent chapters, Jupiter and Saturn experience oppositions and conjunctions to the Sun almost annually—about every 13 and 12.5 months, respectively. Each of these planets is dependable and orderly, paralleling the traditional attributes of their gods. Mars, however, marches to its own drum and takes more than two years—25 or 26 months—to return to an opposition or conjunction, double the time required by Saturn and Jupiter respectively. This isn't erratic, just very Mars-specific. Like a wayward friend on the day you need help moving, Mars shows up when *he* feels like it.

Lest that sound too anthropomorphic, remember that classic depictions of the gods Jupiter and Saturn show mature and stately men, characterizations that carry over to their eponymous planets. Depictions of Mars look like photos from a Sexy Firemen calendar.

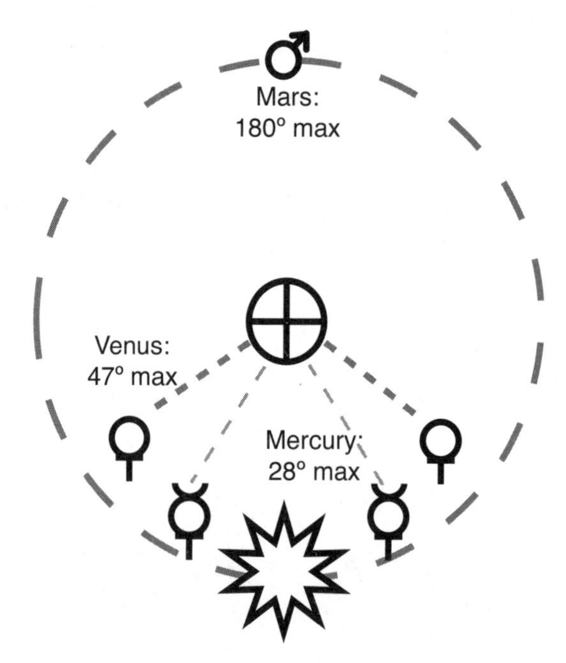

Figure 114. Mars-Sun maximum separation at opposition, expressed astrologically, with Earth as the center point.

Role differentiation by age fits with traditional practices in many cultures. To the Plains Indian tribes, a war chief was generally a younger man, recognized for traits the people respected—honorable behavior, generosity, physical bravery, clever strategy, bold fighting, kindliness, and perhaps charisma. These traits distinguished him in peacetime and could inspire others to follow him into a battle.[102] By contrast, a peace chief was an elder, generally someone who had been a war chief earlier, but who now occupied a less active advisory role.

It's easy to see which role Mars fills, but He was originally called on for protective rather than aggressive combat. Mars was invoked when Rome required protection, then thanked when hostilities were ended—"mustered out," as it were, and returned to his agricultural role in the sensual natural world.

Thanks to His flower-father, from whom we trace the instigating spark, Mars embodied the thrusting-forth energy of the Earth itself. Mars Silvanus was an aspect of this. Offerings were made to this forest deity before fields were cleared for food crops to retain His fertile blessings as humans transitioned a place's plant life from wild to cultivated. Mars: agent of the fertile, with the will to protect this lush garden. Swords into plowshares, and back again, as needed.

Of Heroes, Warriors, and Wildness

Two key words that are frequently used to describe Mars are *warrior* and *passion*.

Often, when we speak of the Warrior archetype, we're speaking of an archetype more like the Hero, better called the *victorious* warrior—someone who has already met and mastered their challenges, a survivor. The terms *Hero* and *Warrior* aren't interchangeable and, when we confuse them, we step into the quandary of trying to glorify every fighter as a hero.

Does a weapon or a combative attitude magically turn me into either a Warrior or a Hero? Hardly. More abstractly, the question speaks to power: power over others versus power from within. When we understand and master (as opposed to conquer) our own skills, drives, and passions, we achieve personal power. When we try to impose mastery over others, look out. Mahatma Gandhi, Martin Luther King Jr., and Mother Teresa are Heroes and can arguably be called Warriors. We shy away from using a term with violent associations to describe them, but we desperately *need* redefinitions.

The term *passion* also demands reexamination. Romance novels? Steamy sexual encounters? Passion is more complex, and includes powerful emotion, compelling desire, exuberant enthusiasm, and hyper-focused energy. None of these terms refer exclusively to romantic love, or even sexual lust. In fact, all of them can apply to any area of life. Passion is my appetite for life. What am I hungry for? How is that hunger whetted, fed, savored, sustained? And how do the passions that I feed in turn sustain me?

The vividness of life can be an indicator of passion. Do you notice color, flavor, sound, aroma, and all the varieties of sensory input that engage the body and awaken the imagination? Or are life's experiences emotionally colorless to you? One beautifully expressive term for the lack of an appetite for life is *ennui,* which "shortens life, and bereaves the day of its light."[103] Byron called ennui "that awful Yawn which Sleep cannot abate."[104] Ennui is boredom, listlessness, an emotional vacuum. It is the heavily sighed "Whatever" and the Sun-bereaved day. When we feel it oozing languorously over us, we should leap up and run like hell.

In shamanic terms, passion is life energy. When it's missing, we're likely to be "porous," vulnerable, not fully present in our own lives. And if I am not fully, vitally "inhabiting" my own skin and psyche, there's room for other energies to move in: germs and illness, self-defeating habits, energy-sucking companions, depression, stuff I'm too susceptible or disengaged to resist. Allegorically, an example of this comes in the classic vampire stories. Dracula can only enter a dwelling if he is invited in.

The antidote to the soul paralysis of ennui is passion—Mars' Green Man energy, the life-force of nature pushing its way up and into the light.

Ironically, culturally, it seems as if what we least want is true exuberant life-force. In *Last Child in the Woods: Saving Our Children from Nature-Deficit Disorder*, Richard Louv writes that, through all of human history, communities have valued their young boys specifically for their strength, natural curiosity, and high energy, which helped them become the effective hunters, defenders, and explorers we needed.

Until now. Youthful energy that found a ready outlet in a range of necessary survival skills now has few outlets at all. In particular, boys who would have been cherished participants in their society because of that terrific energy—budding heroes—are now labeled as having "behavior problems." The diagnoses of Attention Deficit Hyperactivity Disorder (ADHD) have soared for both boys and girls, but 90 percent of those placed on medication are boys.[105] Louv reports that more time spent outside in nature, especially in unstructured play, often moderates ADHD. We divorce ourselves from the Green Man, from Mother Nature, at our own peril.

This divorce from nature plays out in terms of fertility, lust, and passion as well. Much of our food is now "neutered"—created without sexual frisson and carrying no fertile potential. It's been decades since meat animals were left to their own devices for procreation. Their breeding is highly controlled, and artificial insemination has replaced natural mating. Many of our common food plants no longer create their own viable seeds for the following year—the fertility has been engineered out of them—which means that crops can be high-yield, bug- and drought-resistant, and perhaps even delicious, and yet sterile. This is creepy—humans messing with the sex lives of animals and plants. Is it also contagious? The ads for "enhancement" drugs and fertility clinics should make us wonder. We desperately need Mars Silvanus and nature's wild places.

North Germanic Tyr

The Norse deity Tyr is equated with the planet Mars and has some of the same qualities, or at least more of them than any other spare Norse god. Tyr (or Tiwaz, Tui, Zui, and Teiws) appears in the late Icelandic *Edda* writings. His name simply meant "god"; now it appears as Tuesday, literally "Tui's Day," considered the day of Mars. Not many details exist about Tyr, but the story we have about Him and the Great Wolf Fenris is chilling and complex.

An unworldly wolf called Fenris has been created, a child of the trickster god Loki. Predictions have foretold that Fenris will destroy the gods' world, and the creature grows larger and stronger by the day. Since his father is a god, killing Fenris isn't an option, so the other gods decide to bind the beast.

Heavy chains are brought, and through trickery—"We just want to see if you're stronger than these chains"—Fenris allows himself to be shackled.

And he easily breaks free.

And he gets even bigger.

Heavier chains are forged and set with binding spells and mighty locks. "Those first chains were too easy for you. Try these! We're just testing them, mind you…"

And again Fenris quickly snaps the chains—good game!—and continues to grow in size and strength.

Fearing utter destruction if they fail again, the gods commission the dwarves to craft them something special—an enchanted ribbon as light as silk. This ribbon, Gleipnir, is woven from magical ingredients: the beard of a woman, the roots of a mountain, the sensitivity of a bear, the breath of a fish, the spittle of a bird, and the sound of a cat walking—substances no longer found in the world because they were all used up in the crafting of Gleipnir.

The magical ribbon looks innocent, but Fenris—himself a creature of magic—senses something of its mystery. Suspicious of the gods and their binding games, Fenris refuses to be wrapped in the fine soft ribbon unless one of the gods will guarantee their good intentions.

How do you prove yourself to a wolf? By becoming a hostage of sorts: "One of you shall put his hand in my mouth," says Fenris. His proposal is not vicious. He just wants to be dealt with honorably, with a high price paid for any tricks. If everyone acts honorably, there won't be any problems.

But the intentions of the gods aren't honorable, at least not toward Fenris, and they know it. They mean to bind him eternally to prevent him from destroying their world, so whoever takes this pledge will be forsworn and pay the price. No wonder they hang back!

Noble Tyr steps calmly from the hesitant crowd, and places His right hand between Fenris' huge wolf jaws, knowing...

Satisfied by the action of honorable Tyr, Fenris allows himself to be bound yet again. Great game, ha-ha!

Gleipnir is wrapped around him—cat-step, mountain-roots, and all—and then Fenris the wolf tries to free himself.

He cannot.

The gods have played him false, and his great jaws snap shut.

Tyr is renowned for His courage and wisdom, and this tales shows why. He sacrificed Himself twice—first in His honesty, forsworn by the gods' devious intentions. Then He pays the physical penalty, sacrificing His arm, and making no attempt to escape or weasel out of or renegotiate the deal in support of a greater good—saving the world. Being brave when you think you'll emerge unscathed? Big deal. Tyr's action demonstrated bravery in the face of certain sacrifice.

The roots of the word *sacrifice* go back to "sacred" and "ritual," the idea of something being made sacred—*sacri*—by offering it up to the gods. Tyr's action was one of self-sacrifice. Following His encounter with Fenris, Tyr was referred to as "The Leavings of the Wolf," which has a certain grim humor to it, akin to calling Him "Lefty" (see figure 115).

Figure 115. The one-armed god Tyr, following the gods' duplicitous dealing with Fenris. Based on a bronze sculpture from Denmark, c. 200 BCE.

In Tyr, we find physical bravery and the willingness to seek alternatives to warfare. Tyr is credited with inventing the *thing*, the great assembly at which all sides of an issue were aired, all options considered, and requests for judgment brought forth and deliberated. The *thing*—called the *moot* or *folkmoot* in early Anglo-Saxon—was where a community came together to take care of business and, ideally, find diplomatic solutions to any differences. That failing, decisions were reached through champions fighting one-on-one, as an alternative to outright war.

We saw the feminine side of the Tiwaz rune as the spindle in figure 9. As its name implies, the rune is also associated with Tyr—a.k.a. Tiw—for whom it represents a spear. Tyr's noble qualities connect this rune specifically to a spear in the service of justice. The goddess spins the world into creation; Tyr shows us how to act in the world as an inner compass for doing the right thing.

This is Mars as a provocative agent of change and growth, in nature and the green plants, and in the deep zones of our own most passionate connections to life.

Temples—In a Sense

We've seen temples aligned to the Sun, the Moon, and Venus laid across the landscape, and finding a temple physically aligned with Mars would be grand. But Mars' own motion and cycle seem to preclude this. There are plenty of temples dedicated to the *god* Mars, but I found none that actually point at the *planet* Mars—which doesn't mean there aren't any.

But there is a living temple to Mars-as-Tyr: the *thing*. The parliaments of Iceland, Denmark, and Norway all include *thing* in their assemblies' official names; in Sweden, the term is used for local county councils and courts of law. *Thing* also appears in government assembly names for several self-governing territories, like the Isle of Man and the Faroe Islands.

The larger theme is one of people coming together as an assembly to work matters out among themselves—interested parties, experts, wise voices of experience, people who care about the topics. Concerns can be aired and solutions crafted, even if we no longer decide large issues through one-on-one battles between champions—although, as an alternative to war, it's still an appealing idea. Like the peace chiefs of the Plains Indians, this type of gathering is the "peace counsel," people gathering to shape decisions.

An early *thing* meeting site was rediscovered in 2005 in Sherwood Forest. Called Thynghowe on old maps, it was found by local history buffs who spotted an old reference and began exploring. The site is a low, wide mound on a natural low hill, surrounded by oak trees. Ancient markers show that the boundaries of three parishes came together here, making this a geographically equitable place for meetings. Thynghowe may date back to the Bronze Age, and there are other ancient *thing* and *moot* sites throughout the British Isles.

While knowing an actual physical place is great, it's what took place—and what *takes* place—at a *thing* that gives the gathering its power in our imaginations. Totally appropriate to Mars and Tyr, it's the people and their actions that count. The places where people assemble and the living memorial of the people who have acted beyond their personal welfare and sacrificed themselves are what we remember.

Learn by Doing

1. **Cleanse old wounds:** For many of us, our struggles are internal. Remember the people you were passionately attracted to in your youth? How can I live without so-and-so? Remember someone with whom you were passionately furious? How can I live in the same universe as so-and-so? Whether joyous or grim, we carry a lot of that intense energy around with us. As we survive life's trials, we each become the Wolf's Leavings—what the wolf didn't eat. Check the Mars Retrograde chart in appendix C. Then use this planet's lengthy retrograde as a time to do any self-examination and self-healing that you can. See the past clearly and clear the way to move forward unencumbered.

2. **Connect with your power animals and guides:** If you know the practice of shamanic journeying, you can use that as a way to release old conflicts by connecting with your power animals and other guides. Ask to be shown an incident from the past that requires release in the present, then ask for a simple and effective releasing ritual that you can do right away. Stick with key words and phrases like *effective, simple,* and *right away.* And then do it!

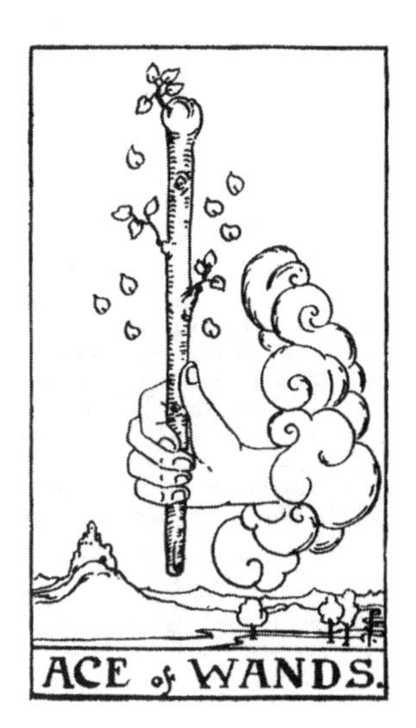

Figure 116. The Ace of Wands.

3. **Explore Mars in the tarot:** For me, in tarot terms, passion relates to the suit of Wands (see figure 116). Where is my passion, my hunger for life? How am I acting (or not acting) on it? If Wands are coming up plentifully in a reading, consider your actions, your passions, and your appetite for life. And be especially willing to notice what may be lacking in these areas.

4. **Get active:** Take the Tyr-Fenris-inspired challenge. The most obvious symbolic choices include working on veterans' issues or directly with wounded vets, and donating to or volunteering in a wolf sanctuary. You can also honor Tyr with other forms of community involvement, like advisory councils and community gardens (also a nod to the Green Man). Let your

activism, whatever it may be, follow the trail of sparks generated by your passions, since that's what gives any of these actions power.

5. **Honor Mars Silvanus:** Go to a place in nature and let your senses open to the spirits there. Take time to sense their presence, and adjust your own tempo. Enjoy and be replenished. Consider leaving a thanks offering. This can be as simple as placing pebbles or fallen leaves in the shape of a heart, or pouring from your water bottle onto a thirsty-looking plant.

Chapter 8

Jupiter—King of Many Names

High in the Rocky Mountains, I have a rendezvous, an intimate encounter, with the King of the Planets.

Amid boulders and sparse pines on a small promontory jutting into a mountain lake, two great blue herons and I watch as the light shifts. The Sun's glow fades in the west-northwest; in the southeast, hints of light from a Full Moon backlight obscuring clouds. The time is twilight—neither light nor dark, but the transition between them—the time of change, of openings between the worlds.

Indeed.

One heron unfolds its great wings and rises into the twilight air. Clearly visible in the soft remains of the light, it flies in a leisurely circle not far above me, long legs at rest, long neck tucked back in its distinctive Z-curve. The bird retraces in the air the circle we've already drawn on the ground for our ritual space. Outside of time and space, the heron moves in slow motion, each wing-beat its own spectacular event. A lake-born breeze that's risen as darkness falls laps the water against the rock-strewn shore, joining the soft tempo of the bird's wings as the only sounds.

The heron repeats its circuit a second time and finds a perch in one of the pines. I realize I've been holding my breath and glance around to get my bearings. The second heron is still attentive at the water's edge, rapier beak and snake-like neck silhouetted in the last residue of sunset light on the water. With a quiet thank-you to the birds, the water, and that moment in time, I turn to go back to camp.

Suddenly, directly ahead, I meet another twilight visitor. In the southeast, above a gleaming bank of cloud, a large and brilliant planet, a Wanderer, emerges. Jupiter. He is bright, close, and dramatic. The land around me seems brighter for his presence, and I freeze in my tracks to watch as he emerges farther from the clouds. I'm caught in his beam—a deer in Zeus' headlights. I've been moonstruck many times, but now I'm Jupiter-struck, Zeus-struck, and I feel infused and giddy with this light. And slightly off-balance as well. Jupiter is *so* bright, *so* big, like a tall stranger stepping from concealment into my path. How did he get so close without me noticing? And how many of Jupiter's stories speak to variations on this theme, as Jupiter-Zeus—He of the high places, with or without disguise—suddenly steps toward a mortal woman?

Gradually, I follow the shadowy path back to drums, friends, and campfire, and a while later, our howls and songs, uproarious and melodious, greet the rising Full Moon. We sing Moon-honoring songs to Her long into the night.

We no longer remember the songs to welcome Him.

The name Jupiter comes from *Dyauspitar,* Sanskrit for "Shining Father." In Latin, this became *Zeuspater,* or "Zeus-Father" (Father Zeus), which eventually shifted to Jupiter. When we speak the name Jupiter, we are speaking the planet's title: "Shining Father."

The ancient peoples knew Jupiter as the King of the Planets. After the Moon and Venus, Jupiter is the biggest and brightest night-sky Wanderer—steady, dependable, huge, and dramatic. Small wonder so many early peoples identified the planet Jupiter with their own idealized ruling deity.

Marduk—The Earliest Jupiter

To the Babylonians, the planet Jupiter was created and named by Marduk, the high god who created the world and established cyclical order in the heavens. The planet was separate from the god, although it symbolized and represented Him, and was generally called Sagmegar or Nibiru. This was the order-keeping watchman of their night sky, the visible expression of Marduk's powers and concerns. The Babylonian creation story, *Enuma Elish,* dates from about 700 BCE, although its roots are far earlier, c. 1800 BCE. Its writer says:

> For the stars of heaven, He upheld the paths . . .
> None among the gods shall transgress Thy boundary . . .
> [Marduk] made the stations for the great gods;
> The stars, their images, as the stars of the Zodiac, He fixed.
> He ordained the year and into sections He divided it . . . [106]

Sure enough, Jupiter's steady motion rarely deviates from the Sun's own path along the backdrop of zodiacal stars, holding to the ecliptical line like its nighttime guardian. He varies at most a mere 1.3° from the ecliptic, moving like a bead sliding along a cord. This is Marduk/Jupiter in his best King of the Planets persona—Establisher of Heavenly Order.

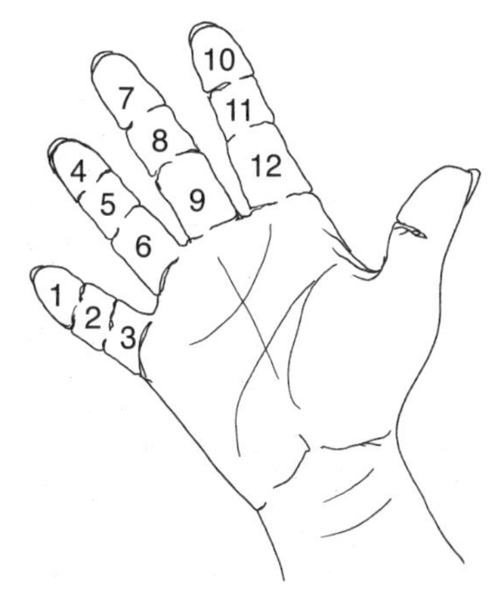

Figure 117. Using finger joints for counting to 12.

His twelve-year planetary cycle was especially important, since the number twelve was key in the sexagesimal (sixty-based) Babylonian number system. This may seem odd to us unless we're buying eggs, but its ancient usage was widespread (see figure 117). Just as the Sun passes through each of the twelve zodiacal constellations in a solar year, Jupiter passes through the same twelve constellations over its twelve-year cycle, returning to the same location every twelve years. Select any Sun-Jupiter conjunction, add twelve years and (generally) five days, and you'll come to the next conjunction in their twelve-year pattern, 5° to 6° farther along in the zodiac than at the conjunction preceding it:

May 13, 2012	Aries-Taurus cusp (astrological: Taurus 23°)
May 18, 2024	Taurus (Taurus 28°)
May 23, 2036	Taurus (Gemini 3°)
May 28, 2048	Taurus (Gemini 8°)

For a more complete listing, see appendix D.

Figure 118. Sagmegar, written in pre-cuneiform symbols.

Jupiter's elegant precision was well-known to the ancients and was responsible for both the planet and the god Jupiter winning descriptors like "steady" and "dependable" even four thousand years ago.

Elsewhere in the *Enuma Elish*, Marduk wields the thunderbolt as his magical tool, removing all impediments to a bountiful, habitable land. Gods associated with Jupiter tend to be big on thunder and lightning. The role of the benevolent Shining Father is expressed in His name, Sagmegar, written in pre-cuneiform pictograms as *Mul* ("planet"), *Sag* ("black-headed people," as the Sumerians called themselves), *Me* ("qualities of civilized life"), and *Gar* ("to place down, to bring or bear").[107] In speaking the name Sagmegar, the Sumerians were saying: "This planet/god brings civilized, well-ordered life down from the heavens to the People" (see figure 118).

Me is the important component here. The *Me* were sacred precepts that signified the attributes of civilization. Their symbol is written as a T, the same shape used for the Greek letter *Tau* and a visual expression of the Golden Mean (see figure 119). *Me*'s presence here speaks to Jupiter's associations with the harmonious aspects of social order, a theme we still connect with the influence of Jupiter.

Figure 119. The Greek letter *Tau*.

Horus—The Egyptian Jupiter

To the Egyptians, the three known outer planets—Mars, Jupiter, and Saturn—were each seen as an aspect of the hawk-headed god Horus. Jupiter-Horus carried a multitude of titles: Horus Who Illuminates the Two Lands, Horus Who Bounds the Two Lands, Light Scatterer, Star of the South, and He Who Opens Mystery.

Although references to the Two Lands are generally assumed to refer to the Upper and Lower Kingdoms of Egypt, this can also refer to Jupiter's steady ecliptical path, the boundary line of northern and southern "lands" in the sky. The planet's hieroglyph shows two arching cobra uraeus figures on the crown, one facing forward and the other back, clearly keeping watch in two directions (see figure 120).

Figure 120. Jupiter's Egyptian hieroglyph.

Zeus—The Greek Jupiter

Zeus had an oracular shrine at Dodona, first dedicated to Great Goddess Dione but later shared with Zeus. Although not as famous as the oracle at Delphi, Dodona was the most ancient Grecian oracle site, in use since about 2000 BCE. Its priestesses and priests lay on the ground among the trees and received prescient knowledge from the rustling of leaves overhead. Rather than announcing what would come, these oracles focused on right action—how best to respond to what the future brought, which is still the most useful type of prophetic information.

Some accounts mention beech trees, but others say oak, the tree most closely associated with Zeus and the planet Jupiter. The oak's great size and strength make this an appropriate plant ally of that bright, large planet and, as the tallest of Old World trees, the oak was also the most susceptible to Zeus' lightning strikes (see figure 121). Largest, highest. Appropriately, the high-soaring, all-seeing eagle is another symbol of Zeus.

One of Zeus' many titles was Zeus Xenios, from *xenos,* meaning "stranger." Hospitality, considered a sacred duty, came under Zeus' purview as well. This benevolent practice had side benefits. News from afar was shared, and reciprocal bonds of honor were established between far-flung communities. But the practice was further elevated as a service to the gods, because "strangers and beggars come in the care of

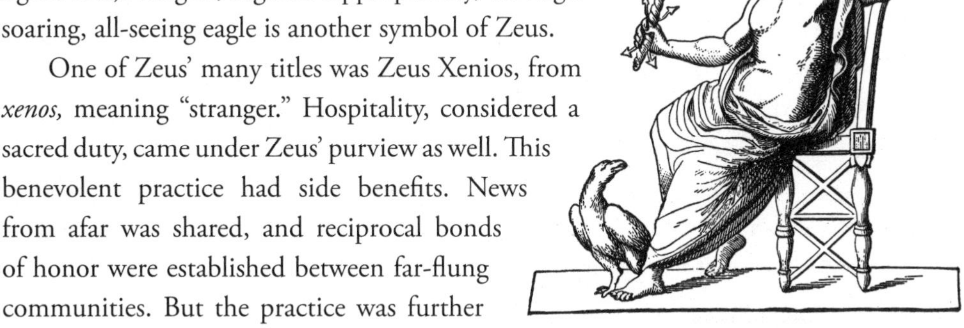

Figure 121. Zeus with lightning bolts, whirlwind (what looks like a twisted baguette), and eagle.

Zeus."[108] This was a matter of ethics as much as etiquette; any stranger could *be* Zeus in disguise. Moreover, hospitality wasn't a uniquely Grecian virtue. For instance, the Celtic lands considered hospitality a sacred responsibility, and a Hindi proverb—*Atithi Devo Bhava*—says "the guest is god."

Our modern interpretations of Zeus read like a celestial soap opera, with Zeus as divine roux and His jealous wife, Hera, plotting revenge on earthly rivals and their half-divine offspring. Zeus has been portrayed as a classy but aging sugar daddy, Mattress King of the Gods. His mortal partners have been described variously as brides, consorts, hostages, and rape victims.

Playboy, sexual predator, or serial rapist? These are expressions of human behavior but are oversimplified in terms of the divine, with the mortal women too often typecast as victims. Surely, real union with a god is more complex.

What if our Shining Father played a more literal part, *in absentia?* Imagine a time when, even if man's procreative role was understood, the role of father wasn't nearly so official—a time when celebratory sexual unions were seasonally enacted to inspire the fertility of crops and animals, when couples ritually came together under the night sky, and when children born of these unions were considered children of the god. This remote Shining Father can be the ideal dad, not bound by the human limitations of an earthly father or the partner-of-the-mother.

That's my personal, earthly theory. Here's a more celestial one: the many guises of Zeus parallel constellations or natural phenomena. Does He dress for the date—cloud banks for Io and meteorites showering Danaë? Some of Zeus' shape-shifting disguises—swan, bull, eagle, snake—can be equated with the night sky's constellations as a storyboard.

The artistic and literary inspirations have continued flowing for centuries, as artists have drawn on this sensual and diverse subject matter, creating nature scenes with animals and meteorological special effects, dramatic subtext, and—naturally—beautiful women, generally looking more enamored than victimized.

Another theory doesn't see the planet Jupiter as Zeus at all. Working with every detail of the text, Florence and Kenneth Wood theorize, in *Homer's Secret Iliad,* that the story entirely parallels celestial motion. References to deities are taken to refer to specific planets, while the epic's mortals are seen as specific stars and constellations. Since He's omnipresent, Zeus isn't portrayed; instead, in the Woods' book, our planet Jupiter is the goddess Athene.[109]

As back-story for the Trojan War, the fair-minded mortal Paris is asked to decide which goddess is the most beautiful—Hera, Athene, or Aphrodite. This is a no-win situation for Paris, since any choice he makes will grievously insult two goddesses, so he succumbs to the best bribe. If Paris declares Her the winner, Aphrodite will give him Helen, the world's most beautiful woman. Hence, the Judgment of Paris. Outcome: love, war, epic storytelling.

In the Woods' premise, Poseidon (Saturn), Ares (Mars), Apollo (Morning Star Mercury), and Hermes (Evening Star Mercury) are among the divine players. Opening the tale are three goddesses, each very beautiful and representing the three brightest objects in the night sky: Aphrodite is the planet Venus; "white-armed" Hera is the Crescent Moon; and Athene is the planet Jupiter. This is a radical shift in how we usually perceive Jupiter—female?!—but remember that Athene was Zeus' daughter, born from His head full-grown and armored for battle. These are the Greek versions of Jupiter and His father-born daughter, Minerva. The young goddess' attributes echo those of Her father: justice, wisdom, industry, and skills in all of the arts and crafts (the civilizing influence), plus the strategies of war. Both She and Zeus carry and use thunderbolts, and both wear the aegis, a protective breastplate.

In *The Iliad*, Hera and Athene are still angry at Aphrodite and Paris about that beauty contest, and often work together against the Trojans. In the Woods' interpretation, this Hera-Athene alliance symbolizes Full Moon conjunctions with Jupiter. These are indeed spectacular—the Moon and Jupiter both directly opposite the Sun and therefore each at its brightest—and these Jupiter-Moon conjunctions happen annually. Athene and full-faced Hera always meet far from Aphrodite, as if shunning her. As we've seen, Venus stays near the Sun and so is *never* near the Full Moon. I've watched this Moon-and-Jupiter versus Venus drama play out in the night sky: it works.

Seeing Homer's masterpiece as an extensive record of the heavens provides an additional motivation for the tale to be told, retold, and handed down so precisely.

Jupiter-Sun Oppositions

Below are the dates when Jupiter is at solar opposition, with the Sun's light hitting it most directly as seen from our earthly vantage point between the two. Find the nearest Full Moons for delicious bright-goddess viewing.

Jupiter is also retrograde at these times, moving slowly—if at all. It may be at a standstill around the dates listed below.

Remember, Jupiter is *visually* in line with these constellations, which is astronomy. It's good for actual night-sky observation. If you're working astrologically, check your ephemeris.

October 29, 2011	Jupiter in Aries
December 3, 2012	Jupiter in Taurus
January 7, 2014	Jupiter in Gemini
February 7, 2015	Jupiter in Cancer
March 8, 2016	Jupiter in Leo
April 8, 2017	Jupiter in Virgo
May 9, 2018	Jupiter in Libra
June 11, 2019	Jupiter in Ophiuchus
July 14, 2020	Jupiter in Sagittarius
August 20, 2021	Jupiter in Capricorn
September 27, 2022	Jupiter in Pisces
November 3, 2023	Jupiter in Aries (this cycle restarts)
December 7, 2024	Jupiter in Taurus
January 10, 2026	Jupiter in Gemini
February 11, 2027	Jupiter in Leo

There's enough in this short list to see the pattern—twelve years plus a few days. A more extensive version is given in appendix D.

Whatever you think of a star-coded *Iliad* or of Jupiter as Athene, the ideas help reestablish that recurring theme. In one myth after another, the gods are portrayed as shape-shifters and tricksters in how they treat each other and us hapless mortals. Whether we meet them in written lore, in art, or in the vast cosmos, they are shape-shifters.

Thor—Jupiter in the North

The Norse myths present another version of our father figure, Zeus. Here His name is Thor, the source of our English word *Thursday,* Thor's day. The German and Dutch words are *Donnerstag* and *donderdag* respectively, meaning

Figure 122. Thor wielding Mjöllnir.

"the Thunderer's day," a respectful work-around, rather than referring to the god by name. Thursday was sacred to Thor. Records note that a 7th-century priest scolded his backsliding parishioners for honoring the old ways on Thursdays.[110]

Son of Odin and the giantess Jord, whose name meant "Earth," Thor was a thunder god, with a thunder-and-lightning-creating hammer named Mjöllnir as His main weapon, tool, and symbol (see figure 122).

The *Edda,* a 13th-century Icelandic manuscript derived from earlier lore, tells the stories of gods and heroes. One of these tales, "The Lay of Thymyr," is intriguing for its theme of disguise: Thor must masquerade as the goddess Freyja. The giant Thymyr stole Mjöllnir; Thor gets His hammer back if Freyja marries Thymyr. The "bride" duly appears—Thor tricked out in veils—and mayhem ensues. This is all patently absurd, of course—imagine Zeus disguised as Aphrodite—but let's go a bit deeper.

Thor is identified with the planet Jupiter, and Freyja with Venus. Of all the planets, these are the two that are consistently mistaken for each other. Venus is *never* aloft at midnight, always staying closer to the Sun. Jupiter isn't as bright as Venus, but is still brighter than every other planet. Both have a clear white color, while Mars has an orange tone and Saturn leans toward cool light blue or soft dull yellow, depending on the angle of its rings. Is the Jupiter-Venus confusion an undercurrent of this poem?

This would seem more far-fetched, perhaps, if the two planets rarely met in the sky for comparison. Surprise! They get together on a regular basis. Appendix D gives only their closest encounters in the coming years.

Can Jupiter and Venus be mistaken for each other as Thor was when disguised as Freyja? You be the judge.

Add in an occasional Crescent Moon as white-armed Hera, and we see the same trio of goddesses that created such a dilemma for Paris.

Figure 123. Saami depiction of Thor and His hammer, from the design on a drum pre-1830.

The Saami people of Lapland and other animistic and shamanic groups in Northern Europe paid homage to Thor. The sound of the drum is akin to thunder, and their frame drums were decorated with images of Thor (also called Thoragales and Horagalles) and other deities in the Lower, Middle, and Upper shamanic realms (see figure 123).[111] Many such drums have dangling bits of metal attached behind the head, adding to the crash-boom effect.

When Christian missionaries arrived in these areas, it was the Saami's drums—the magical tools that helped the people connect with their gods—that the Church actively destroyed. They ultimately were not successful, as the Saami shamanic culture is still active.

Druids and the Oak

The oak tree is among the sacred trees symbolized in the Celtic Ogham alphabet; it is expressed as *Duir,* with the phonetic value of the letter D (see figure 124).[112] The oak was considered a noble tree, a chieftain tree, and the Druids had ceremonial uses for every part of the plant—twig, bough, branch, bark, acorn, and leaf. *Duir* relates to the word for "door" in a range of languages—*doras* (Irish), *thura* (Greek), and *Tür* (German)—as does the letter/sound D in other languages—*daleth* (Hebrew, where it symbolizes both "door" and "authority") and *delta* (Greek, where it also means "fourfold," as in the four elements and material-world wholeness). Fine doors are still made of oak, for strength. In Ogham, *Duir* symbolized strength in a magical sense, representing one who encompassed the strength or power to access the portals between the worlds. To use that door as an entrance is to en-trance, to commune with the divine. Doors and boundaries give—and limit—access to these other realms, like Jupiter moving vigilantly along the ecliptic between the worlds as the powerful, generous, dependable, and thunder-wielding King of the Planets.

Figure 124. *Duir,* in the Ogham alphabet.

Learn by Doing

1. **Find Jupiter in the night sky:** Since Jupiter moves slowly, once you've spotted him, you can locate him easily on subsequent evenings, at roughly the same time in nearly the same place as the weeks unfold.

2. **Connect with the Shining Father:** This can be especially important and poignant if your relationship with your own father is problematic or absent. For me, this has meant gradually shifting into a heart place of seeing Jupiter and knowing that the benevolent father energies of the universe are watching out for me.

3. **Learn to drum:** Does Mjöllnir speak to you as a symbol of Thor, of thunder, of a valued tool? Hammering, building, blacksmithing. How about the thundering sound of a drum? If you've never done any drumming, this is a vast exploration in itself. Whether you discover the joys of the drum through an ecstatic drum circle (rhythmic, melodic, transporting, danceable), or through shamanic practices (the drumbeat as the "vehicle" for passage to and from the spirit realms), this form of "thunder" plays a vital role in spiritual practices worldwide.

4. **Welcome the unknown wanderer:** How do we treat those who are strangers to us? Whether a store clerk, a panhandler, or another driver, what thoughts and actions do they receive from us? How many of them are Zeus, Thor, or Athene in disguise? Although we may not trace our personal lineage back to a Jupiter-lit grove, what if we just decide we *are* children of the gods and goddesses—all of us—and treat each other accordingly? More personally, how can we each find an inner connection to deity at those moments when we're feeling like strangers to ourselves, and need to direct some divine hospitality toward ourselves?

5. **Keep a journal or meditate:** What were you doing twelve years ago? Especially consider the time during Jupiter's lengthy retrograde. Are you harboring any unresolved baggage from that time, stuff you'd rather not carry into the future? Were there dreams then that are finally ready to take flight? You can look at this through journaling or as a meditation (whether guided or silent).

6. **Try a Jupiter tarot layout:** If tarot is a good tool for you, explore Jupiter's energies via a reading. Don't concern yourself with the astrological place of Jupiter in relation to your chart, as that isn't needed here. It's the time

frame of your life experiences that you'll be exploring.[113] What was taking place for you twelve years ago? Jupiter engages in an annual multi-month retrograde, and that apparent backward motion is a good time to reflect back on your own journeys over the last twelve years. What were your interests, intellectual pursuits, projects, and activities? What was your

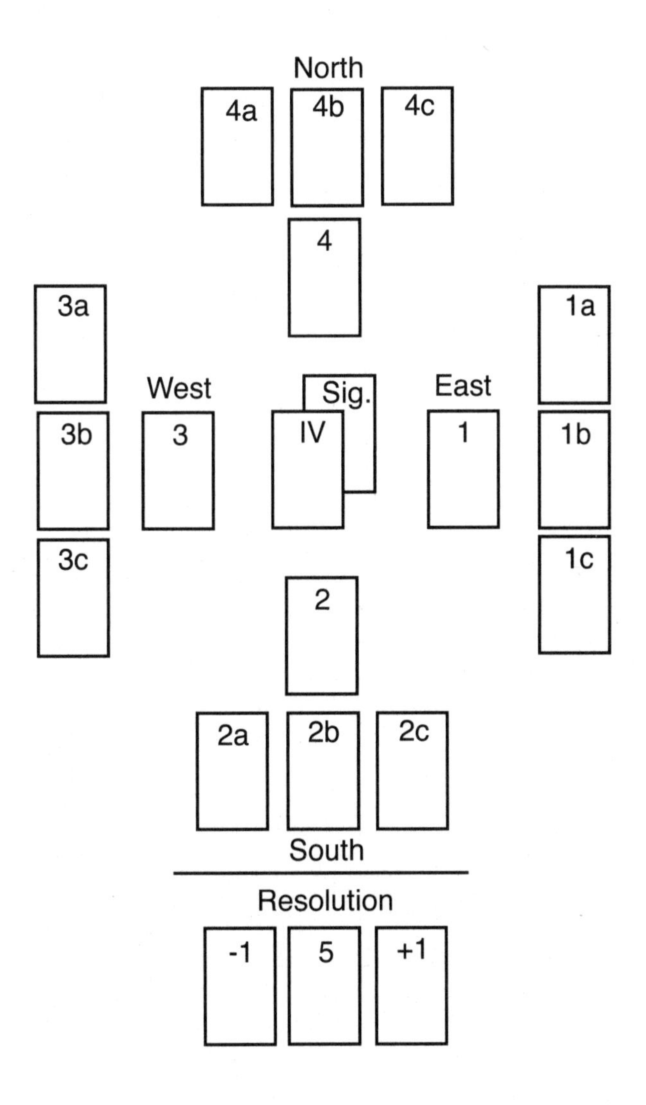

Figure 125. Jupiter tarot layout.

emotional life like in terms of romance and other significant matters of the heart? In terms of livelihood and finances, what was your situation at the time?

Chances are you'll see amazing growth, but you may also find that some old baggage has followed you into the present. Strategies that worked well then may have lost their usefulness but may still be part of your *modus operandi* simply out of habit. Jupiter symbolizes expansion and the bountiful energy of increase—getting even more of what you've already got. You want that celestial growth to work on what's good in your life rather than to increase any issues still lingering. The tarot spread helps you look back, then points at ways of releasing, resolving, and removing any old concerns that have followed you into the present. Jupiter will be back right here again in another twelve years. Find a means of moving on *now,* so these particular problematic aspects don't follow you for another twelve years.

1. In the center position, place Major Arcana IV, the Emperor, to represent the planet Jupiter (see figure 126). (If your astrology-tarot system differs, use the card you identify as Jupiter here, as long as the card has positive associations.)

2. Select a Court Card as Significator, and position it slightly behind your Jupiter card.

3. Shuffle the remaining cards well, then deal the four top cards into positions 1, 2, 3, and 4. These represent:

   East—the ideas that engaged you twelve years ago
   South—the activities you were involved in twelve years ago
   West—your emotional life twelve years ago
   North—your career/financial state twelve years ago

   Explore each of these cards, interpreting and taking notes before proceeding. Work with intuitive impressions as well as standard meanings.

4. Now pull another three cards for each direction (a-b-c), to represent your journey on each theme over the last twelve years—a three-card snapshot of

lessons offered in each area, learned or not learned, matters resolved or still dangling. Take notes.

5. If needed, select a card from among the a-b-c cards that represents some past challenge still unresolved. Perhaps there are several, but concentrate on the one that feels most problematic. This becomes card 5.

6. Put this card, representing unresolved business, back in the remaining deck and shuffle anew, concentrating on your desire for resolution. Then go through the deck until you find that same card and place it in position 5 again. Take the neighboring cards—those on either side of it in the deck—and place them in the bracketing positions (-1 and +1). These represent practical steps you can take now to resolve the lingering issue.

7. Follow through! Take the actions necessary to bring resolution in these matters.

Figure 126. Major Arcana card IV, the Emperor.

Chapter 9

Saturn—The Ancients' Final Frontier

Long ago, before Greece existed, before Time existed, there was an earth goddess named Gaia and a sky god named Ouranos. Each night, Ouranos covered Earth and, as a result, Gaia bore many children. Jealous of the young Titans He had engendered, Ouranos hid some of them deep within the Earth, where their discomfort caused horrible pain to their mother, Gaia. At last, the youngest child, Cronus, escaped this fate and grew to manhood.

Longing for relief, Mother Gaia pinned all Her hopes on Cronus. She crafted a sickle—the crescent-bladed knife used to prune vines and harvest grain—and gave it to Cronus, along with grim instructions for its use (see figure 127). The sickle was swung and—chop!—Cronus cut off His father-god's genitals. After that, the sky stayed far away from the earth.

Cronus became the new king. He married His sister Rhea, and together they brought forth a new generation of divine beings. But—fate repeating itself—Cronus lived in fear of the prophecy that one of His children would take His place. Rather than risk this, Cronus swallowed each child when it was born.

Figure 127. Cronus, a.k.a. Saturn.

Finally, one child, Zeus, was hidden from His rapacious father and grew to manhood. Through trickery, Zeus caused Cronus to drink an emetic potion, and soon Cronus began to vomit up all the other children. Being gods, they were unharmed.

Cronus was deposed, conquered. Zeus banished Him as far away as possible—He chained Him at the edge of the universe.

So runs the sad, strange story of Cronus, the Greek Titan the Romans knew as Saturn. The metaphors in this shudder-inducing tale relate to life, death, and mortality. The sky-father Ouranos must ultimately answer to Earth (Gaia), who rules the realm of physical reality. At the behest of Gaia, Cronus (Saturn) is put in charge of imposing limits. When limits are imposed on *Him,* however, Saturn finds Himself imprisoned at the outer edge of space, which is where we still can find Him as the last planet visible to the naked eye.

Without the dramatic lore or incredible brightness of Venus and Jupiter, without the unusual color of Mars, without the curiosity-provoking speed and elusiveness of Mercury, and without being able to see unassisted the rings that are now Saturn's claim to fame (first spotted by Galileo in 1610), we have only old, sparse Saturn references on which we can draw. All of these reinforce Saturn's symbolic role as one of *finality.*

Slowly, slowly, the planet Saturn moves against the backdrop of the zodiac. His deliberate journey embodies both the comprehension and the control of time, moving along the boundary between being and non-being. Until Uranus' discovery in 1781, Saturn was the outer-most known planet, the sky's final moving object before the starry backdrop, the "firmament."[114] Thus the belief that, in seeing Saturn, you're seeing the edge of space before reaching the stars themselves. Well, that's final.

There are some layers here of similar names and themes between Cronus and Chronos, two distinct characters who gradually overlapped. The familiar depiction of Chronos is as Father Time, complete with a long white beard, an hourglass, and a scythe (the full-scale version of the sickle)—all symbolic attributes of advancing age, passing time, and death's finality. Father Time looks like a benign hermit; his more sinister interpretation is as the Grim Reaper. The theme, however, is the same: we're only here for a while. Chronos has also been depicted as an elderly man holding the zodiacal wheel, as if controlling time in a full-circle, what-goes-around-comes-around way.

Chronos means "time"; the term shows up now in words like *chrono*graph (time piece), dendro*chrono*logy (time measured via tree-rings), and syn*chron*icity (apparently unrelated events united by their timing). Neoplatonic philosopher and grammarian Macrobius (395–423 CE) wrote of Cronus (Saturn) *being* Chronos (Time).[115] Crone is another unrelated word, but the similarity in sound and theme resonate anyway.

As the outermost and slowest moving of the visible planets, Saturn is unique in ways that are reflected in this planet's lore. In Zeus-Jupiter, we met a benevolent ruler—stately, wise, considered, steady—moving through the nearly twelve-year Jupiter cycle. By contrast, Saturn's cycle is far longer—nearly thirty years (29.46 years, which is rounded to 29.5 years). The Moon sails quickly through the zodiac, averaging just 2.25 days in each sign; Saturn spends an average 2.5 *years* in each sign. His is a slow and deliberate journey that embodies both time and life.

Through thousands of years and many generations, thirty years was the average human life span—a statistic that held true from roughly 30,000 through about 2000 BCE. Saturn's life-span-like cycle playing out against the zodiacal backdrop was gradually noted. Thirty years became a recognizable and noteworthy span of time. Egyptian pharaohs who held the throne that long enjoyed special jubilee celebrations during the thirtieth year of their reign, as did Persian Shahs.[116] One of the Sanskrit names for the planet Saturn is *Kala,* which means both "time" and "death," as well as "blue-black," and that color evokes images of Kali dancing on the outer edge of dark space, draped in Her necklace of skulls.[117]

The themes inherent in Saturn's lore are shape and boundaries, limitations, time, and finality. These expressions do not convey simply a droning of doom-and-gloom, but rather acknowledge the larger life-rhythm of completion: everything that begins will eventually end.

Saturn, so remote and slow, tends to earn attention in historical astronomical writing only by getting together with other planets. Meetings with Jupiter are especially noteworthy because they are relatively rare, since Saturn and Jupiter are the slowest of the visible planets. Like a cross-country family reunion for elderly relatives who hate traveling, when Saturn and Jupiter make the trip and get together, it's a big deal. In their alter-egos as Cronus and Zeus, with their miserable shared history, they're obviously loath to meet often. Early astrologers interpreted their meeting accordingly—as portents of big things to come.

Sometimes, depending on their mutual retrogrades, Jupiter and Saturn meet *repeatedly* over several months. This happens every couple of hundred years, most recently in 1940–1941, and next in 2239 CE. And it happened, very notably, two thousand years ago.

The "Star" of Bethlehem?

Barely visible in the western sky at sunset, Saturn and Jupiter were drawing together, within 10° of each other, on February 10, 7 BCE. Then both disappeared into the sunset as they each conjuncted the Sun.

By the Spring Equinox, on March 23, both planets were reappearing as Morning Stars, with all the rebirth imagery attendant on Morning Star appearances. Close together, less than 2° apart, from mid-May to mid-June, Saturn and Jupiter were exactly conjunct on May 28.

They moved apart slightly—not far, not even 3°—and then both began their retrograde motion. Astrologer-astronomers, already watching, would *really* have been paying attention by this time. Jupiter, the faster of the two even in reverse, closed the gap. They were closest from September 26 to October 4, both at a standstill with less than 1° of separation, united as one heavenly phenomenon. Or one "star"? On which day were they closest together? Exact conjunction dates have been calculated (several different ones are put forth), but at the time, to the observer, both planets simply stood still. This was a multi-day phenomenon. We miss that—wonder, alarm, awe?—when we try to express this event as an exact single date.

The two planets edged apart, but remained near each other into December—closest again from December 2 to December 12, at 1°03' (another *multi-day* occurrence)—but even as both resumed direct motion, they were still just 5° apart in early February of 6 BCE.

During the previous conjunction in 26 BCE, Jupiter and Saturn approached, met, and separated within a mere three months. The 7–6 BCE meeting, as Saturn a[illegible]piter lingered so near each other for months, was really exceptional. A[illegible]ortents of big things to come.

The two slow wanderers meet every twenty years. Their next conjunction is on December 20, 2020, on the Winter Solstice, when they will be so close together that they may look like a single light in the sky. Look for them in the west at sundown, bright and obvious thanks to Jupiter's presence. Start watching in November to see them move into position; by January, the Sun will overtake them, and they will vanish into the sunset. They'll reappear around the Spring Equinox as Morning Stars, as in 7 BCE, but without the phenomena of extra conjunctions. More future dates are given in appendix E.

Saturnalia—Eat, Drink, and Be Merry

One of the best festivals to mark endings was the Roman Saturnalia, a celebration surrounding the Winter Solstice. This festival, a city-wide break from Rome's strict roles and regimentation, went on from December 17 through the 23. The few emperors foolish enough to suggest shortening the celebration met such serious resistance that they backed down. The roots of how we celebrate both Christmas and New Year's Eve are based in the Saturnalia: gifts, parties, feasting, gifts, funny hats, feasting, parties...

The event sported Saturn's name because of the theme: the end. The Sun and the year had both reached their limits; to celebrate, limits of many other kinds were (briefly) banished. Masters served a formal meal to their slaves, and powerful citizens abandoned their standard-business-garb togas for informal attire.

Whatever your usual form of dress, it changed during Saturnalia. Everyone wore the red *pileus,* a floppy conical hat. Picture the red feather crest on a pileated woodpecker's head *à la* Woody the Woodpecker. That's the *pileus* shape, traditionally worn by freed Roman slaves. At Saturnalia, it was part of the role-reversal costume, signaling a mass declaration of freedom.

When we alter our standard garb and drop our usual sense of self—and especially if we can't be recognized—we tend to behave differently, maybe even to misbehave. Try imagining your city's entire population celebrating in Halloween costumes—for a *week*—to get an idea of the Saturnalia's scope and potential wackiness.

In addition to servant-master role-swapping dinners, there were other festivities: lots of visiting, feasting, partying, and gift-giving. Nowadays, the word

Saturnalian conjures up truly rowdy images (i.e., Bacchanalian), but "taken to the limits" would be a more Saturn-appropriate phrase, with an eye to making way for the new.

Those stereotypical images we use for New Year's Eve—the diapered baby and scythe-toting Father Time—have their roots in the Saturnalia as well. The baby is the new year, reborn at Winter Solstice as the returning Sun. Chronos/Father Time is Saturn, harvesting away the old year with his scythe.

Sugar Skulls and Marigolds

Chronos' scythe brings us back to death. *Día de los Muertos*—the Day of the Dead—is a Mexican holiday/holy day now celebrated in many cities in the United States. The celebrations look superficially like Halloween, taking place on nearly the same date (November 2) and with similar trappings (candy, skeletons). But they are significantly different—if not from a traditional ancestor-honoring Samhain, at least from the costumed keg party that Halloween has become.

On Día de los Muertos, you visit the graves of your loved ones and perhaps decorate them with candles and marigolds, autumn-blooming flowers native to Mexico and Central America, and traditional for this time. When the grave site is tidied, you lay out a picnic of the deceased's favorite food and drink. You may create an altar in your home, full of photographs and significant objects that remind you vividly of the person. Actively remembering them is vital. Families reminisce about those who have crossed over, and children hear stories about relatives they never knew. This day, this night, they are present in spirit, near you again. So you dish up a plate of food for them, pour them a drink, sing their favorite songs.

The overt acknowledgment of mortality is an important component of the celebrations, blended with macabre humor. Three-dimensional skulls made of sugar or chocolate are a traditional Día de los Muertos treat, with the recipient's name written in icing across the brow. These are given as gifts to both the living and the dead to signify the unspoken message: Eventually we're all just bones.

Día de los Muertos celebrations stand in radical contrast to how the subject of death was treated—with hushed tones and euphemisms—during my childhood. How could you *not* be afraid of death if grown-ups were so intimidated by this mystery?

Día de los Muertos shifts away that dread. Children have an annual cultural experience that speaks of death, but playfully and as a natural part of life. The *act* of dying may still be painful, tragic, and profoundly mournful. But Death? That's an altar, a picnic, a piñata for the kids, and the evocative scent of marigolds. Rather than the Grim Reaper—chop!—here you meet Catrina, a fashionable skeleton in a huge *fin de siècle* hat, who offers you a little sugar skull. Your name is on it, and she's smiling. It's a skeletal rictus grin, but she's smiling.

We all depart eventually, and fear isn't a good ally for the transition.

God of the Harvest

Scythes, sickles. We can see Cronus and His cutting tools clearly in the context of the harvest. Cronus can double as Sabazius, God of the Barley, who was honored and then ritually "killed" (harvested), as in the old lay about John Barleycorn.[118] In this traditional harvest song, the grain personifies a man who is cut down but eventually springs back up, returning to "life" as grain alcohol. Phrases like "cut down in their prime" relate back to images of mowing.

With Saturn or Cronus, it's the cutting down that we focus on. Harvest marks an end to the growing season, the end of a plant's life. Robert Graves puts this ceremonial harvesting in the seventh month. In the old Roman calendar, when the year officially began in March, the seventh month was September: *septem* means "seven." This is appropriate for the planet Saturn, always listed seventh in the old Chaldean planetary order, in which there are seven planets, with Saturn as the last—at the very end. Of course, Saturday—Saturn's day—is the seventh day of the week, so again we see the theme of endings.

The Chaldean Planetary Order

Metaphysical planetary uses are largely based on the "Chaldean order," a perception of planetary orbits and proximity attributed to the Babylonians. Watching the Moon, Mercury, and Venus from our Earth-based perspective, the ancients saw that all three sometimes moved between the Earth and the Sun. Mercury and Venus also go behind the Sun, but the Moon never does, so the Moon must be the closest to Earth. Mars, Jupiter, and Saturn never come between the Earth and the Sun, but they do move behind the Sun in their circles, so they must be farthest away (see figures 128 and 129). This is how the Sun comes to be in the center of this arrangement, based on which planets move in front of it and which never do. The Chaldean order is incorrect, but not foolish, and it remained in metaphysical use long after people came to understand that the Earth orbits the Sun. And we use it still. Magical squares are based on the Chaldean order, as are most other numerical associations with the planets.

The Saturn Return

Through the thousands of years in which the average human life span was thirty years, humans had plenty of time to notice the apparent synchronicity in the heavens. Saturn and its 29.5-year orbit became encumbered with a huge load of prophetic baggage. Saturn returns to the astrological sign it occupied at your birth and—chop!—time's up. Once upon a time, and for a very long time, that prophecy had fabulous odds of being accurate.

Figure 128. The Chaldean order of planets: Moon, Mercury, Venus, Sun, Mars, Jupiter, and Saturn, all circling the Earth.

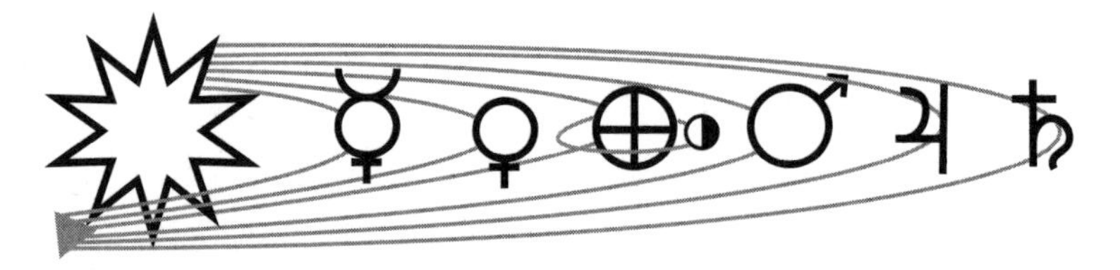

Figure 129. The correct order of the planets, moving out from the Sun, Mercury through Saturn.

Even though the average life span is now longer, that Saturn anxiety is deeply embedded in our psyches. Saturn returns are referred to in mainstream conversation almost as frequently as Mercury retrogrades—and with considerably more dread, as if the Grim Reaper were literally and purposely approaching your door. People may not expect literal death at that time, but often brace themselves for upheaval, chaos, and general karmic butt-kicking. The power of self-fulfilling prophecy has a way of obliging these dismal predictions, but Saturn is blameless.

So let's reframe the Saturn return like this: Things that mattered before can be reviewed and reexplored now from a vastly expanded perspective. Perhaps we'll face fears, doubts, and isolation as part of the journey. But confronting and moving through these impediments—or chimeras—will give us the new tools we need for whatever comes next.

Saturn's Double Theme

Throughout this chapter, there have actually been two themes contrasting, conflicting, and intertwining: "We perceive limits" and "We burst through limits."

We tend to imagine iron shackles when the word *limits* appears. But the word encompasses many things: shape, edges, boundaries, form, definition, limits established, limits challenged, limits questioned and discarded, limits as "good boundaries," limits we need to exceed, limits due for re-evaluation, and limits that we establish to give ourselves personal shape, form, and definition of identity, without which we'd be amorphous.

Figure 130. The weighing scales of Thoth and Maat—a human heart in the left pan, Maat's feather in the right.

Saturn—Cronus, plant-fertility god and son of Gaia—symbolically sparks our drive to push forward, while simultaneously holding the sickle. As if carefully pruning a treasured vine, we use the blade to trim away what is dead and finished, what can be left behind.

In the lore of Egypt, Thoth weighs the heart of every person after they die, placing it on a great scale with a feather in the other pan that represents the goddess Maat (see figure 130). Your heart should be lighter than Her feather, unencumbered by bitterness, meanness, and hatred, and ready for your journey to the Other Side.

We trim and pare and move forward.

Creating a Saturn Altar

Although there were temples dedicated to the god Saturn, they weren't aligned to the motion of the planet Saturn. His place was the sky, the edge of the universe. But what if you need a focal point around which to rally Saturn energies—a place, a location?

The Día de los Muertos–style ancestor altar can fill this role. It can be elaborate, with places to stash genealogical research and all your photo albums. Or it can be incredibly simple—a leaf from your great-grandfather's cherry tree, an index card with a recipe in your grandmother's handwriting. I have a jar of soil and small stones from my childhood yard, my grandmothers' yards, and other places important to me. This represents ties to my own past, and deeper earthy and earthly roots back through my ancestors.

These are physical forms, appropriate for physical-form human beings; but what about an outer-edge sky god?

Saturn's token is the sickle, a tool that whittles away the old beliefs that limit us, scratches a line in the soil and defines our space, and finally, ultimately, snips the fine-spun thread of life that holds us to the Earth—like Atropos and her shears—when we reach the end of our measured time. Shears, or a sickle, can be appropriate on this altar.

But Saturn's true temple is the sky. Whatever props and reminders we may create indoors, his realm is out beyond the walls.

Learn by Doing

1. **Make a will:** This is a good idea, especially if you have children or a partner, or own property. Also consider the practical measure of enlisting a trusted person to make decisions about your care if you are unable to do so. This is variously called a Health Care Power of Attorney, or Power of Attorney with Durable Provisions. Along the same lines, there is a simple form called "Five Wishes" that, when signed and witnessed, covers many of the same concerns.[119] Whenever and however we depart, lessening the confusion among our care providers by having some clear steps already marked out will help smooth the way.

2. **Explore your limits:** Where do you feel blocked on your path through life? Conversely, is your path so ill-defined that you keep losing it? Speaking metaphorically, do you need a snowplow, or curbs and banisters? Limits can slam you to a halt, or—as form, shape, and definition—work in your favor. Figure out what kind of limits you need, or need to get rid of, and reshape your parameters, as if with careful sickle-blade whittling.

3. **Make a Möbius strip:** The lemniscate that we saw earlier, the symbol of infinity, can apply to Saturn, but with a twist. As the sky and the ultimate edge, Saturn is infinite; in our own dealings, in human form, we are finite. This is quietly expressed three-dimensionally by the Möbius strip—a strip of paper that has two different sides that run together.

4. **Stop and smell the marigolds:** Even if you're likely to live more than thirty years, this human experience is relatively brief. Regardless of any unresolved issues with individual ancestors, you're the product of a long line of amazing survivors. Does your niece's laugh remind you of your late brother's? Does your child's expression look like your feisty grandmother's? Here you are, on Earth, in human form. Hug your friends and smell the marigolds. Use Día de los Muertos, Samhain, or a similar occasion to remember your ancestors and say thanks.

Chapter 10

Some Special Stars, Groups, and Phenomena

We've looked at the circumpolar stars—present and past—at the constellations of the zodiac, and at the planets that move along that starlit ecliptical path. Now we'll look more specifically at some other stars that have attracted human attention and imagination. Some are zodiacal, but distinguished on their own; others are far from the zodiac path, but so bright or noteworthy or significantly placed that they command interest.

Bighorn Medicine Wheel, Wyoming, USA (44° 49' N, 107° 55' W; 1200–1700 CE)

"Medicine wheel" stone arrangements are as impressive as modern observatories, and they mark a wide range of stars. This particular site consists entirely of small rocks—lines of them, laid out like the spokes of a wheel high in the Bighorn Range, a spur of the Rocky Mountains (see figure 131). At an altitude of over 9600 feet (2926 m), the site is remote, with open 360° views and changeable weather. Snow in June is entirely possible.

The wheel itself is about 80 feet in diameter (24.38 m), with a center cairn about 2 feet high (0.6 m). There are twenty-eight spokes, plus other rock cairns at varying points around the rim. These provide alignments to the Summer Solstice and to the brightest stars that first herald its approach, then count off the days following it.[120] The Winter Solstice and the equinoxes aren't marked; the weather here makes autumn, winter, and spring use unlikely.

Around 1200 CE, the Summer Solstice took place on June 14. Leading up to that date, the bright star Fomalhaut (in Piscis Austrinus) had its heliacal rising. Fomalhaut is far south of the ecliptical path of the Sun and the planets. It's visible from this panoramic vista, but it never rises even a hand span above the southern horizon as seen from the Bighorn latitude. Fomalhaut's timing is its special attraction, specifically its heliacal appearance. Its brief pre-dawn blink in mid-May came twenty-eight days before the solstice (see figure 132).

About twenty-six days later, far to the north of Fomalhaut's rising point, Taurus' Aldebaran made its heliacal appearance. At that time, Aldebaran's heliacal rising was before the solstice and served as its herald. The solstice came a

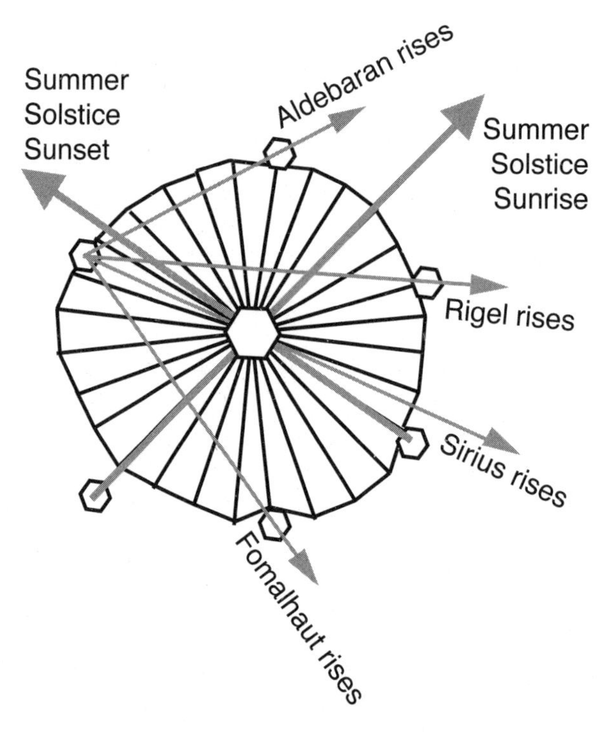

Figure 131. Bighorn Medicine Wheel, aerial view. Aligned through the rock cairns, each thin arrow points to the location of a star's heliacal rising. Thick arrows show the rising or setting point of the Sun at Summer Solstice.

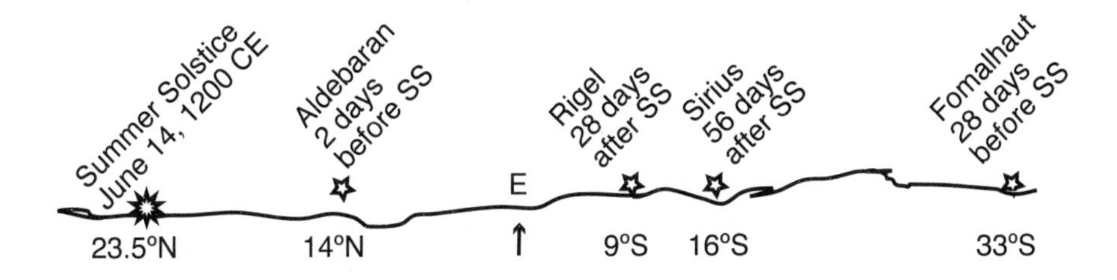

Figure 132. The horizon array of the Bighorn Medicine Wheel's aligned stars, and the Summer Solstice (SS).

couple days later, with both rising and setting marked by Bighorn Medicine Wheel alignments.

Next, the days *after* the solstice are counted out on the wheel: Orion's star Rigel has its heliacal appearance twenty-eight days after the Summer Solstice (mid-July), followed in another twenty-eight days by the heliacal appearance of Canis Major's Sirius (early August).

There are perhaps 150 medicine wheels in the northern Rocky Mountain region, stretching up into Canada and out into the outlying ranges. Far to the south, outside Tucson, Arizona, a similar wheel at Zodiac Ridge marks different celestial positions. The Zodiac Ridge wheel is also round and has cairn markers, but no spokes. Weather isn't an issue here, so both solstices are marked, along with the equinoxes and lunar maximum extremes. And more stars: Polaris is marked, along with the heliacal setting of Auriga's Capella at 45° N in the northwest (early June), and the heliacal risings of Rigel, Virgo's Spica (7° S; mid-October), and Lyra's Vega (38° N; mid-November), as if continuing the steady count found for the summer months at the Bighorn Medicine Wheel.[121] The people who created Zodiac Ridge were settled, farming people with the advantage of stable, year-round viewing to make note of lunar extremes, which wouldn't be as obvious to nomadic people.

Fomalhaut, Capella, Aldebaran, Rigel, Sirius, Spica, Vega—these are some of the brightest stars and among the first visible in the evening sky. Notably, they are scattered throughout the sky, often far off the ecliptic.

Vega is high overhead on the meridian at 9 p.m. in early September. To find it, locate Sagittarius' Teapot asterism and then look straight up (see figures 28 and 30).

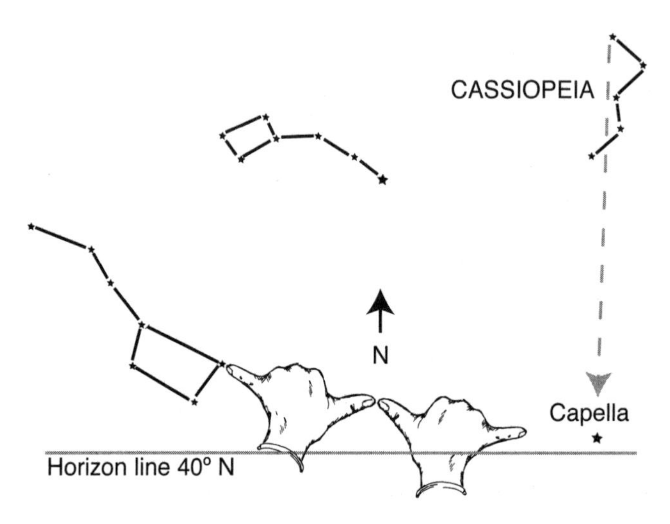

Figure 133. Capella in the north-northeast, early September around 10 p.m.

Fomalhaut is below Aquarius, always low in the southern sky, but quite bright (see figure 31). Find it low along the meridian at midnight around the time of the Autumn Equinox.

"Little she-goat" Capella can be located via the Big Dipper, or you can find distinctive Cassiopeia, another circumpolar constellation, and work from there (see figure 133). It's a bright autumn star, rising in the north-northeast sky around 10 p.m. in early September, then rising earlier and riding higher each evening as the weeks pass.

You can see what a vast expanse of sky these stars occupy. All they have in common is magnitude, or brightness. A quartz crystal was recovered along the equinox lines at Zodiac Ridge, naturally six-sided but described as "worn." Intriguingly, crystals are used in Navajo healing work. Some "singers," Navajo healing practitioners, look through a quartz crystal toward first-magnitude stars and use their refracted colors to determine the appropriate ceremony for a client.[122]

The Milky Way—World Tree and Path into Eternity

The Milky Way is one of the most beautiful and least seen of the sky's wonders. This vast veil of stars stretches through the entire sky in a soft-edged band. It's easily quenched by city lights, and it hides from moonlight as well. Predictably, milk figures in many of its myths. Hercules suckled on Hera, but her milk spilled. In Egypt, this celestial milk came from the cow goddess Hathor. *Gala* is Greek for "milk," and we are part of the redundantly named Milky Way *gala*xy.

The Milky Way cuts through the zodiac at Scorpio and Sagittarius on one side and at Taurus and Gemini on the other. With the ecliptical band of the zodiac encircling us horizontally and the Milky Way looping over and under us vertically (albeit not exactly through the poles), we see the quartered circle of the medicine wheel expressed three dimensionally (see figure 134). In other words, we aren't just "on" the wheel noting the four directions, we are also "within" it.

To the Maya, the Milky Way had several identities, which changed as the Earth's rotation shifted the Milky Way's position. It was seen as the World Tree with a crocodile head at its base when thrusting up from the Earth along the southern horizon between Scorpio and Sagittarius; it became the Cosmic Monster when arced from east to west, and a canoe paddled by gods and animals that sank as the Earth turned.[123] The strange (to our eyes) illustrations that embellish every ancient Maya site are pictorial sky maps, at least in some interpretations.[124]

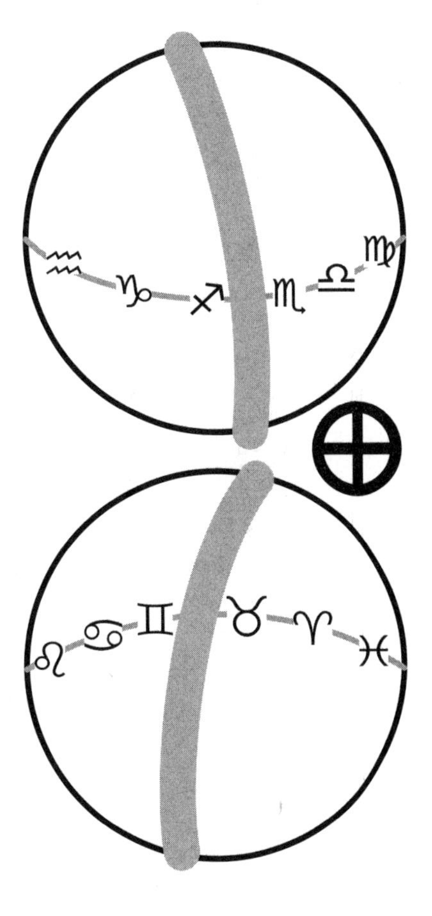

Figure 134. The path of the Milky Way transecting the zodiac.

The Milky Way also figures into 2012 theories, which have evolved into the "War of the Worlds" of our time—Orson Welles' 1938 radio-play based on H. G. Wells' Martian invasion story. Unfortunately, some who tuned in mid-broadcast thought they were listening to a live newscast and were terrified, although the stories of widespread panic in the streets are themselves overstated. Twenty-twelve has become the new version of this—a chance to be totally fear-addled if you choose. The fearful response isn't mandatory, nor is it a healthful contribution to our collective thinking. The Twenty-twelve myth has its roots loosely entwined in Maya lore based on their mythic date of creation—August 11, 3114 BCE—carried forward to December 21, 2012, the Winter Solstice. Some interpretations leap off into apocalyptic conclusions about the end of the world. Remember the ancestral apprehension about the Winter Solstice? "Will the Sun return from the south?" This is the modern version of it—on steroids.

What *will* happen in 2012? The precession of the equinoxes has gradually shifted the Sun's location at the Winter Solstice into line with the so-called galactic equator, an *imaginary* line that runs through the center of the Milky Way galaxy. On the 2012 Winter Solstice, the Sun will purportedly rise most precisely where the ecliptic and the galactic equator cross (see figure 135). The Sun *already* crosses this same vague point annually, and has done so for aeons.The Winter Solstice Sun has been brushing along this point for about thirty years, and will continue to do so for another thirty. Remember, once the Sun rises, no stars are visible, so actually *seeing* this isn't an option.

According to astronomer and Maya expert Anthony Aveni, there are no direct mentions of the Milky Way in Maya inscriptions, nor any identified glyphs meaning Milky Way, nor any evidence that the ancient Maya cared about the precession.[125] The few Maya texts that refer to the 2012 date don't portray it as cataclysmic, simply as the end of an age *as based on their calendar system*—just as New Year's Eve in 1999 was for us, based on *our* calendar system, despite all the nerve-wracking predictions of Y2K disasters. Respond as seems best to you, with pragmatic measures or a lively Saturnalia.

Higher up in figure 135 is constellation Cygnus as we see it—as a long-necked swan flying downward. Others have envisioned a long-tailed bird flying upward. Many cultures saw the Milky Way as a path into the Afterlife, the world of departed spirits, complete with a bird spirit to assist them on the journey. Cygnus may be that bird, and variations abound: swan-maidens in folktales, cranes in Celtic lore, the Water Bird symbol in the Native American Church, the vulture carvings at Göbekli Tepe. Even the dove as Holy Spirit may trace back to this significant constellation (see figure 136). This bird spirit—bird *as* spirit, bird as spiritual communicator with the Next World—flies between the worlds along the Milky Way's road, a celestial path for the spirits of the dead.

Drawn on the Land?

A group of cathedrals in France dedicated to *Notre-Dame* (which simply means "our lady") are believed to have been intentionally laid out across the landscape in the pattern of Virgo's main stars. The arrangement—if that is indeed what it is—stretches across many miles but is reasonably accurate when matched with

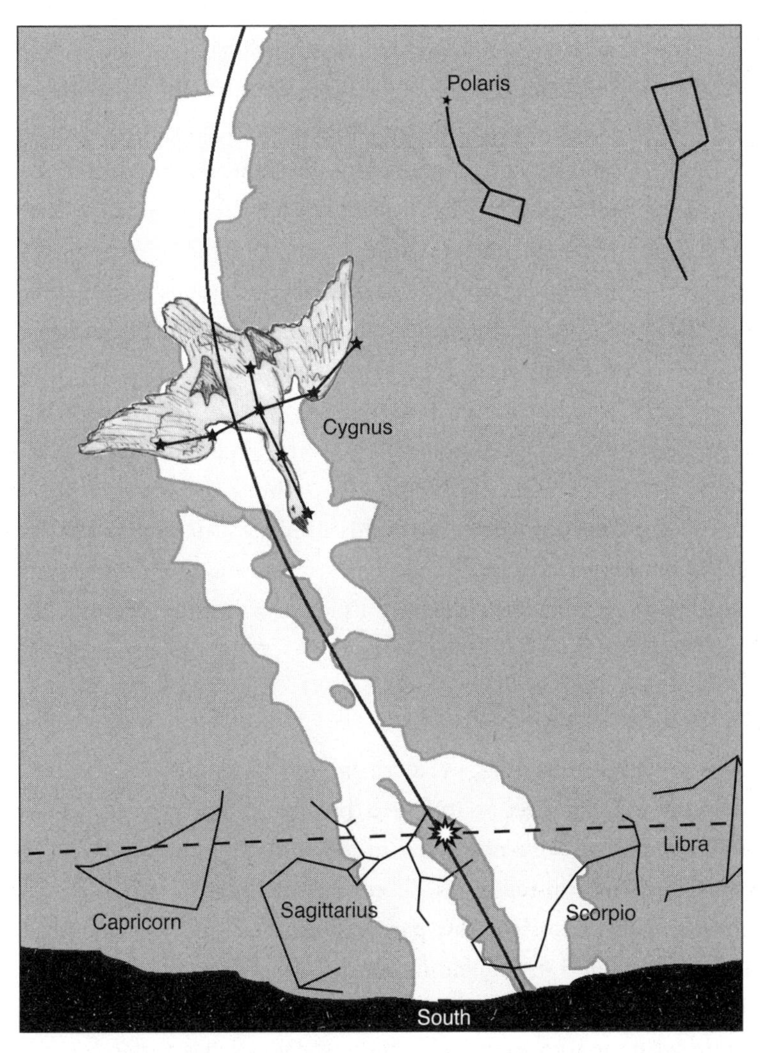

Figure 135. The ecliptic intersecting the galactic equator region amid the Milky Way, with the Sun in its 2012 Winter Solstice position.

an image of Virgo (see figure 137). Many of these cathedrals are built atop old goddess sites, as if to invoke the grain goddess to protect and care for an agricultural region.[126]

Figure 136. The Ace of Cups, with its descending dove.

Nor is this the only place where star-based arrangements may appear. Theory holds that the fields around Glastonbury form shapes symbolizing the zodiac signs and other constellations.[127] The aerial photographs are interesting. The small, odd-shaped fields can be combined any which way, so—yes—images emerge. Elvis, anyone? My objections are aesthetic. The earth- and stoneworks seen throughout this book are graceful, carefully wrought, and elegant in line. But these Glastonbury "figures" look forced and contorted, lacking the precision and functional beauty of the other sites. Are these just odd-shaped fields in weird groupings?

Another theory posits that the Pueblo Indian villages throughout a wide expanse of the American Southwest are laid out in the star pattern of Orion.[128] As with the Virgo cathedrals, this involves a very large area and a dozen or so sites. But, unlike the Glastonbury zodiac, there's a sense of precision here. It sounds unlikely, but remember the buildings in Chaco Canyon are aligned to lunar extremes and to each other in a great, interconnected web. So why not here? This idea isn't so very different.

Biting into Halley's Comet?

Earlier, we saw Serpent Mound, the earth-mound effigy in southern Ohio. Theories have proposed a connection to Draco, but that's a northern star group, while Serpent Mound aligns west-northwest to the Summer Solstice sunset. There are other snake/serpent constellations, of course, including Hydra. This monster stretches over 100° across the sky, parallel to the ecliptic, with its head to the west below Cancer and its tail ending much farther east under Libra (see figure 138). It's big, it's a snake, and it's visible from Serpent Mound. But is there anything more substantial to connect Hydra to the great snake effigy?

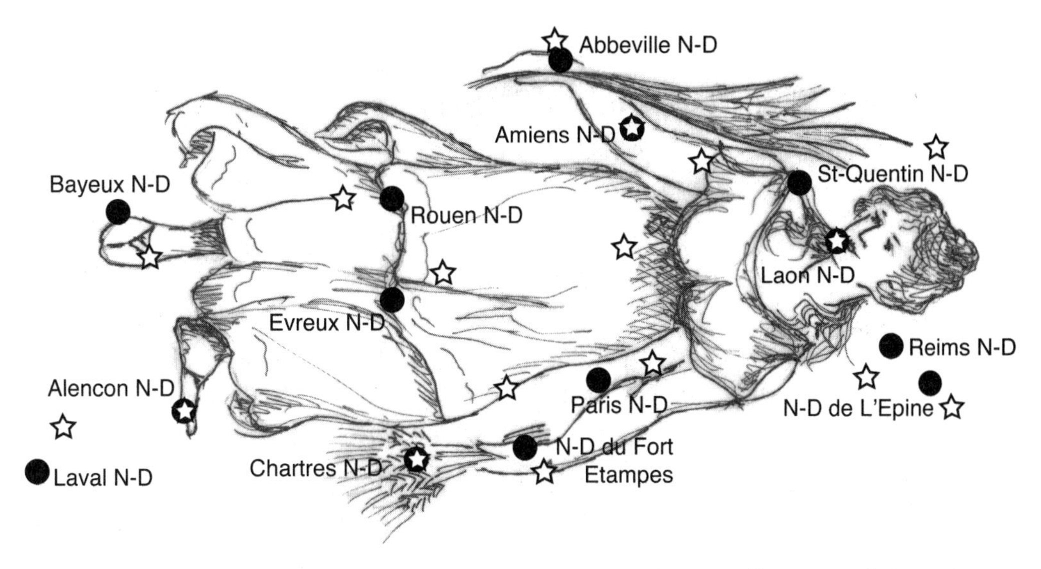

Figure 137. Cathedral sites are shown as black dots, superimposed over an illustration of Virgo. The cathedrals in Alencon, Amiens, Laon, and Chartres appear to be most accurately sited.

The estimated date for Serpent Mound is 1070 CE, plus or minus seventy years. Halley's Comet made a notable appearance in 1066 (and reappears every seventy-six years), so there is speculation based just on those dates that the mound relates to the comet. But how, precisely? Serpent Mound has a long, undulating tail; comets have long, straight tails.

What if the serpent's *egg* is the comet? Is this a record of where Halley's Comet appeared? Setting my astronomical software to 1066 CE, I found Halley's Comet in Canis Minor, which I verified with a local planetarium. Only the constellation's name, Canis Minor, was displayed during our mutual research.

Then I turned on the "illustration" option in my software and found Halley's Comet sitting—in Canis Minor—just in front of Hydra's mouth, like the egg in the mouth of the Serpent.

Tracking comet locations from ancient notes and images is speculative, but with what's known now about its route through the sky, it seems that Halley's Comet was traversing the western end of Hydra for a portion of each appearance since at least 607 CE, gradually edging farther west across the head and out in front of the snake's nose into Canis Minor.

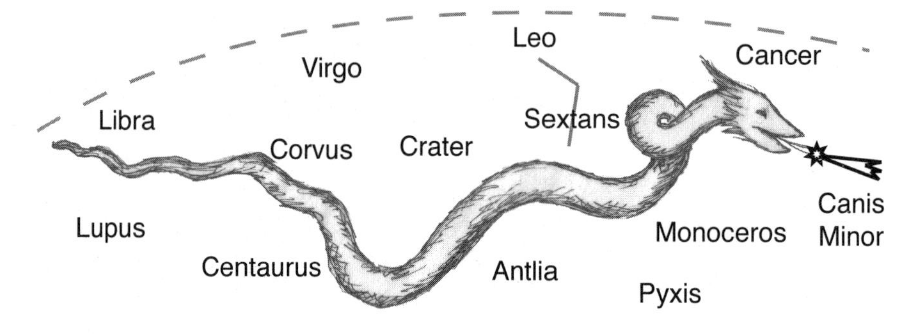

Figure 138. Halley's Comet in 1066 CE, positioned near the mouth of Hydra.

There's a significant problem with this theory, however. We have no idea if the builders of the Serpent Mound effigy saw the stars of Hydra in the same way we do, as a snake—which makes this theory just delicious food for thought.

Orion and His Lore-Laden Belt

When it comes to identifiable constellations, many people cite Orion's belt as the stars they recognize most easily and watch for most often. The belt is the precisely spaced group of three stars across the "waist" of Orion's hourglass-shaped torso. The three smaller stars that dangle below the belt are his sword or its scabbard (see figure 139).

Even people who can't locate the Big Dipper can often find Orion, or least his famous three-star belt. This constellation is visible in the winter months; in fact, an hour after the setting Sun vanishes in the southwest at the time of the Winter Solstice, you can watch Orion rising due east. By midnight, he's high overhead and impossible to miss. The constellation gradually moves west through the winter and spring, finally setting just behind the Sun in mid-May, when he is lost in the Sun's glare. Orion reappears in July, around mid-month.

The Greek tales say Orion, the world's most handsome mortal man, was indiscreet with the Pleiadian sister Merope; in punishment, her father put out Orion's eyes. An oracle directed the blind man to journey toward the east and direct his sightless gaze to the point where Sun god Helios first arose from the ocean. Blind Orion grabbed a guide to lead him and began his quest. The

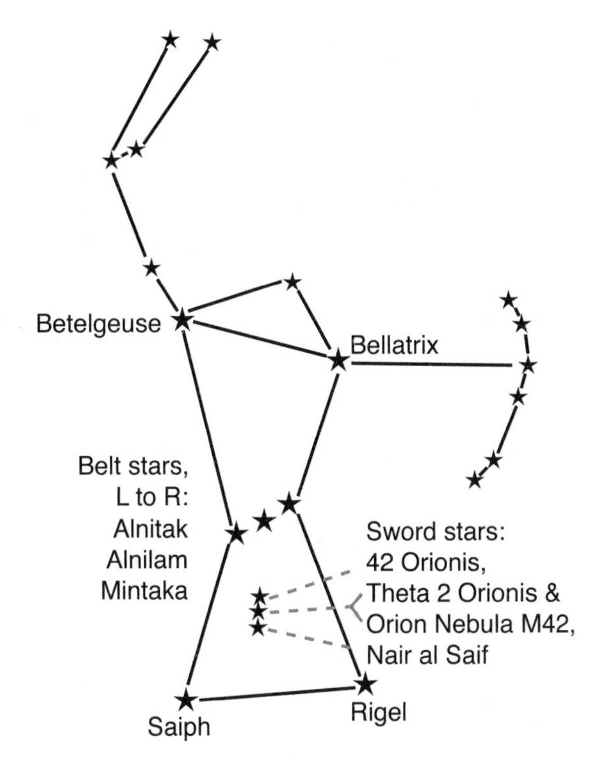

Figure 139. Orion the Hunter shown full-length, with key stars named.

prophecy was fulfilled, and his sight returned. He promptly fell in love with the goddess Eos, but soon returned to his pursuit of the Pleiadian sisterhood.[129]

Put another way, the constellation disappears into the Sun's glare at sundown (Orion's "blindness") as the Sun approaches Taurus. Orion has its heliacal reappearance about two months later (after his guide leads "blind" Orion through the Sun-conjunct darkness into the east), when his sight (or our sight *of* him) is restored. Orion's romance with Eos begins immediately, since Her name means "dawn." She and the handsome mortal are together for a while in the pre-sunrise sky. Then, as the Sun and seasons shift, Orion "deserts" Eos, rising farther ahead of the Sun with each passing month. He's back to his pursuit of the Pleiades, who always rise before him and are always, eternally, 30° ahead as he chases them through the night sky. Eventually, he'll be blinded again.

Or not. For this isn't the only Orion myth.

To the Maya, Orion isn't a man or a god, but a symbolic location—*Oxib' Xk'ub'*, or "Three Hearthstones," representing the main stones that form the traditional Quiché Maya kitchen hearth. If you've cooked out rough over a campfire, you know that three equidistant, equal-sized stones make a stable, triangulated base for your cookware. To the Quiché, Orion's sword-stars are the fire's embers, and the slightly blurred visual effect of the center sword-star—actually a rich group of stars that includes the Orion Nebula—is seen as the fire's smoke (see figure 140). Going back to the very beginning of the world, the first creation of the Maya gods was a central hearth.[130]

The Aztec knew Orion's belt as a bow drill for making fire, calling on the same fiery theme, but from a different perspective (see figure 141).[131]

To the Scandinavian people, Orion's belt and sword were Freyja's distaff. We saw something of the lore of spinning and weaving in the North Star chapter, and we reencounter that theme here, with some important twists. While spindles are associated with Frigg and her realm of domestic concerns, the distaff is connected with the goddess Freyja, who rules the less homey versions of love, beauty, sexuality, and fertility, as well as magic and the transformations of life and death. Frigg's North Star spindle stays in one spot, while Freyja's distaff traverses the sky from east to west. Spinning spells can fall under the heading of domestic magic, but the magical workings associated with the distaff take on official stature.

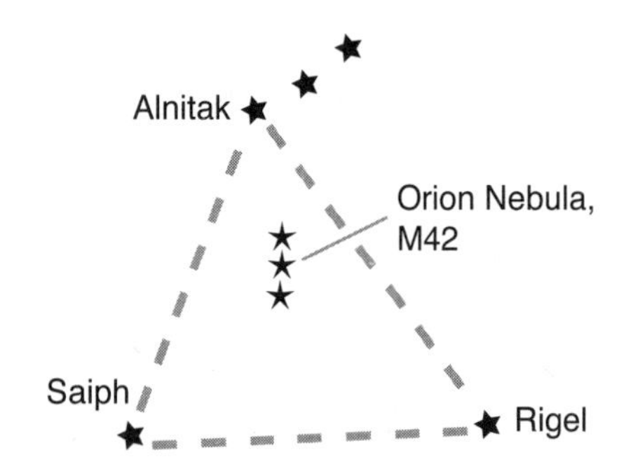

Figure 140. Orion as a cooking hearth.

Figure 141. Orion's belt and sword as a bow-style fire drill.

The distaff holds the spinner's un-spun fibers neatly, ready at hand. Cleaned, combed, smoothed, but not yet twisted into yarn, this material was looped around the "basket" portion of the staff (see figure 142).

The distaff—which is a staff, after all—takes its name from *dís*, old Norse for "lady." This is a lady's staff. But *dís* and its plural *dísir* can also refer to sisters, maidens, female ancestral guardian spirits, and fate-wielding goddesses.[132] Like jewelry, weapons, and other objects of significance, the distaff's quality matched its owner's station. The simplest version is a three-pronged branch (see figure 143), but, like many items, the distaff evolved with ornamental variations and was carved, personalized, and improved upon.

Figure 142. Frigg and her handmaiden, Fulla, with a "dressed" distaff behind them.

Older artworks have shown witches riding what look like backward brooms, but these are probably distaffs. This metaphorical expression of the völva—with her staff of office as a token of her journey into a prophetic trance state—gradually became our stereotypical depiction of a traveling witch.

In the hands of a völva, the distaff symbolized access to the spirit realm. Among the early Scandinavian people, the völva—shaman, seer, oracle, seidhr-worker, and most often female—was a highly respected personage, and the seidhstafr—the völva's distaff—was her staff of office.

Wooden or metal, these distaffs were often ornamented with gems and carvings, but otherwise closely resembled household distaffs. Iron distaffs are sometimes found in the graves of women. Too heavy to function as a practical piece of spinning equipment, these metal distaffs are believed to be seidhstafrs, and the graves are thought to be those of völva.

The seidhstafr didn't hold fibers. It held potential. It was a token that—through trance work, in the shamanic journey, and perhaps by using this special staff to aim and send energy—reality could be manipulated like fibrous strands and woven like cloth.[133]

Völva ("magic-wielding soothsayeress"[134]), *gydhjar* (priestess or goddess), *spákona* (clairvoyant, prophetess)—these terms are all feminine, and the women who practiced these callings were well-respected and well compensated.[135] When the prophetess Thorbjorg arrives in the *Saga of Eric the Red*, she receives warm and respectful welcome and a seat in an honored position, complete with a feather-stuffed cushion. "A staff she had in her hand, with a knob thereon; it was ornamented with brass, and inlaid with gems round about the knob."[136] That "knob" is the jazzed-up version of the distaff's basket. Other accounts speak of a völva traveling with an entourage of assistants, young men and women who will sing her into her trance state.

After she rested from her travels and enjoyed an evening or two of feasting, the völva would exercise her powers within a ceremony. These descriptions read like a Drawing Down the Moon ritual in modern witchcraft traditions, or like transfiguration in some modern shamanic practices, in which the personal identity is put aside and connection with the divine is welcomed, accepted, and often perceptible to others who are present.[137] Prophetic messages may be delivered, warnings given, and danger hopefully averted. Sponsoring a völva and her retinue/support team for several days was expensive—like bringing in a consultant—as was compensating her with appropriate gifts when she was ready to depart.

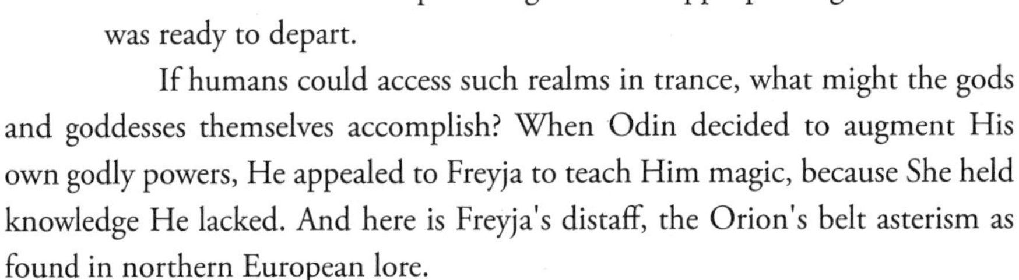

Figure 143. The simplest stick distaff, with bent twigs for the basket, "undressed."

If humans could access such realms in trance, what might the gods and goddesses themselves accomplish? When Odin decided to augment His own godly powers, He appealed to Freyja to teach Him magic, because She held knowledge He lacked. And here is Freyja's distaff, the Orion's belt asterism as found in northern European lore.

Which brings us back to magic and the night sky—full of legendary beings and objects, goddesses and gods, healers and tricksters. The sky was—and is—alight with shape-shifting deities and tools that work magic when handled properly. Planetary motion plays out our largest human themes: birth and death and rebirth; journeys into Eternity; rivalries, challenges, loyal companionship, and noble behavior; even the creation of the world itself. Our building projects have incorporated and marked the sky's rhythms as eloquent reminders of what our earliest ancestors valued—a vivid and lively connection with the spirit world that is given imagery, dimension, and duration by the vast dome of the sky.

Our modern connections with nature have become tenuous, even in the daylight hours, and are nearly nonexistent after dark. We cite loss of open space, poor air quality, nighttime light pollution, and crime rates; but the fact is that nature didn't leave us. Gradually, so gradually, we left it. We remember—sort of—as in a delicious spirit-nurturing dream that lingers at the edge of memory as we wake. But then we forget, and our spirit hungers for manifestations of the sacred, for revelatory experiences. And for beauty and simple wonder.

It's time to go outside and greet the night sky.

Appendix A

Mercury Elongations 2010–2050

Here are Mercury's elongations for the period 2010 to 2050. E means Mercury is east (left) of the Sun; W means Mercury is west (right) of the Sun. The higher the number just in front of the degree symbol (°), the farther from the Sun Mercury will be, improving your chance of spotting it.

Mercury's Morning Elongations

Visible up to about ninety minutes before dawn; check your local sunrise time, find a wide and unobstructed view to the east, and get out there early. Remember, the rooster is another of Mercury's associates in the animal kingdom. This is why.

2010

Jan 27—24.8° W
May 26—25.1° W
Sep 19—17.9° W

2011

Jan 9—23.3° W
May 7—26.6° W
Sep 3—18.1° W
Dec 23—21.8° W

2012
Apr 18—27.5° W
Aug 16—18.7° W
Dec 4—20.6° W

2013
Mar 31—27.8° W
Jul 30—19.6° W
Nov 18—19.5° W

2014
Mar 14—27.6° W
Jul 12—20.9° W
Nov 1—18.7° W

2015
Feb 24—26.7° W
Jun 24—22.5° W
Oct 16—18.1° W

2016
Feb 7—25.6° W
June 5—24.2° W
Sept 28—17.9° W

2017
Jan 19—24.1° W
May 17—25.8° W
Sept 12—17.9° W

2018
Jan 1—22.7° W
Apr 29—27.0° W
Aug 26—18.3° W
Dec 15—21.3° W

2019
Apr 11—27.7° W
Aug 9—19.0° W
Nov 28—20.1° W

2020
Mar 24—27.8° W
July 22—20.1° W
Nov 10—19.1° W

2021
Mar 6—27.3° W
July 4—21.6° W
Oct 25—18.4° W

2022
Feb 16—26.3° W
June 16—23.2° W
Oct 8—18.0° W

2023
Jan 30—25.0° W
May 29—24.9° W
Sept 22—17.9° W

2024
Jan 12—23.5° W
May 9—26.4° W
Sept 5—18.1° W
Dec 25—22.0° W

2025
Apr 21—27.4° W
Aug 19—18.6° W
Dec 7—20.7° W

2026
Apr 3—27.8° W
Aug 2—19.5° W
Nov 20—19.6° W

2027
Mar 17—27.6° W
July 15—20.7° W
Nov 4—18.8° W

2028
Feb 27—26.9° W
June 26—22.2° W
Oct 17—18.2° W

2029
Feb 9—25.7° W
June 8—23.9° W
Oct 1—17.9° W

2030
Jan 22—24.4° W
May 21—25.6° W
Sept 15—17.9° W

2031
Jan 4—22.9° W
May 2—26.9° W
Aug 29—18.2° W
Dec 18—21.5° W

2032
Apr 13—27.7° W
Aug 11—18.9° W
Nov 30—20.2° W

2033
Mar 27—27.8° W
July 25—20.0° W
Nov 13—19.2° W

2034
Mar 9—27.4° W
July 7—21.3° W
Oct 28—18.5° W

2035
Feb 19—26.4° W
June 19—22.9° W
Oct 11—18.0° W

2036
Feb 2—25.2° W
May 31—24.6° W
Sept 24—17.9° W

2037
Jan 14—23.7° W
May 12—26.2° W
Sept 7—18.0° W
Dec 28—22.3° W

2038
Apr 24—27.3° W
Aug 22—18.5° W
Dec 10—20.9° W

2039
Apr 6—27.8° W
Aug 5—19.3° W
Nov 23—19.8° W

2040
Mar 19—27.7° W
July 17—20.5° W
Nov 6—18.9° W

2041
Mar 1—27.0° W
June 29—22.0° W
Oct 20—18.3° W

2042
Feb 12—25.9° W
Jun 11—23.7° W
Oct 4—17.9° W

2043
Jan 25—24.6° W
May 24—25.3° W
Sep 18—17.9° W

2044
Jan 7—23.1° W
May 4—26.7° W
Aug 31—18.2° W
Dec 20—21.7° W

2045
Apr 16—27.6° W
Aug 14—18.8° W
Dec 3—20.4° W

2046
Mar 30—27.8° W
Jul 28—19.8° W
Nov 16—19.4° W

2047
Mar 12—27.5° W
Jul 10—21.1° W
Oct 30—18.6° W

2048
Feb 22—26.6° W
Jun 21—22.7° W
Oct 13—18.1° W

2049
Feb 4—25.4° W
Jun 3—24.4° W
Sep 27—17.9° W

2050
Jan 17—23.9° W
May 16—26.0° W
Sep 10—18.0° W
Dec 31—22.5° W

Mercury's Sunset Elongations

2010
Apr 8—19.3° E
Aug 7—27.4° E
Dec 1—21.5° E

2011
Mar 23—18.6° E
Jul 20—26.8° E
Nov 14—22.7° E

2012
Mar 5—18.2° E
Jul 1—25.7° E
Oct 26—24.1° E

2013
Feb 16—18.1° E
Jun 12—24.3° E
Oct 9—25.3° E

2014
Jan 31—18.4° E
May 25—22.7° E
Sep 21—26.4° E

2015
Jan 14—18.9° E
May 7—21.2° E
Sep 4—27.1° E
Dec 29—19.7° E

2016
Apr 18—19.9° E
Aug 16—27.4° E
Dec 11—20.8° E

2017
Apr 1—19.0° E
July 30—27.2° E
Nov 24—22.0° E

2018
Mar 15—18.4° E
July 12—26.4° E
Nov 6—23.3° E

2019
Feb 27—18.1° E
June 23—25.2° E
Oct 20—24.6° E

2020
Feb 10—18.2° E
June 4—23.6° E
Oct 1—25.8° E

2021
Jan 24—18.6° E
May 17—22.0° E
Sept 14—26.8° E

2022
Jan 7—19.2° E
Apr 29—20.6° E
Aug 27—27.3° E
Dec 21—20.1° E

2023
Apr 11—19.5° E
Aug 10—27.4° E
Dec 4—21.3° E

2024
Mar 24—18.7° E
July 22—26.9° E
Nov 16—22.5° E

2025
Mar 8—18.2° E
July 4—25.9° E
Oct 29—23.9° E

2026
Feb 19—18.1° E
June 15—24.5° E
Oct 12—25.2° E

2027
Feb 3—18.3° E
May 28—22.9° E
Sept 24—26.3° E

2028
Jan 17—18.8° E
May 9—21.4° E
Sept 6—27.1° E
Dec 31—19.6° E

2029
Apr 21—20.1° E
Aug 19—27.4° E
Dec 14—20.6° E

2030
Apr 4—19.1° E
Aug 2—27.3° E
Nov 26—21.8° E

2031
Mar 18—18.5° E
July 15—26.6° E
Nov 9—23.1° E

2032
Feb 29—18.2° E
June 26—25.4° E
Oct 22—24.4° E

2033
Feb 12—18.2° E
June 7—23.8° E
Oct 4—25.7° E

2034
Jan 26—18.5° E
May 20—22.2° E
Sept 17—26.6° E

2035
Jan 10—19.1° E
May 2—20.8° E
Aug 30—27.3° E
Dec 24—20.0° E

2036
Apr 13—19.6° E
Aug 12—27.4° E
Dec 6—21.1° E

2037
Mar 27—18.8° E
July 25—27.0° E
Nov 19—22.4° E

2038
Mar 11—18.3° E
July 7—26.1° E
Nov 1—23.7° E

2039
Feb 22—18.1° E
June 18—24.7° E
Oct 15—25.0° E

2040
Feb 6—18.3° E
May 30—23.2° E
Sept 26—26.1° E

2041
Jan 19—18.7° E
May 12—21.6° E
Sept 9—27.0° E

2042
Jan 2—19.5° E
Apr 24—20.3° E
Aug 22—27.4° E
Dec 17—20.4° E

2043
Apr 7—19.2° E
Aug 5—27.3° E
Nov 29—21.6° E

2044
Mar 20—18.5° E
Jul 17—26.7° E
Nov 11—22.9° E

2045
Mar 3—18.2° E
Jun 29—25.6° E
Oct 25—24.2° E

2046
Feb 15—18.1° E
Jun 10—24.1° E
Oct 7—25.5° E

2047
Jan 29—18.4° E
May 23—22.5° E
Sep 20—26.5° E

2048
Jan 13—19.0° E
May 4—21.0° E
Sep 1—27.2° E
Dec 26—19.8° E

2049
Apr 16—19.8° E
Aug 15—27.4° E
Dec 9—20.9° E

2050
Mar 30—18.9° E
Jul 28—27.1° E
Nov 22—22.2° E[1]

---

1. Thanks to John Walker's public domain site, Mercury Chaser's Calculator, for this information: *www.fourmilab.ch/images/3planets/elongation.html*

Appendix B

Venus Elongations 2010–2050

Here are dates and sky locations for Venus' greatest elongations, her widest separations from the Sun. E means Venus is east (left) of the Sun; W means Venus is west (right) of the Sun. The zodiac constellations are those you'll see in the sky with Venus, astronomical rather than astrological. Any accompanying planets are listed as well. Remember, visibility isn't only for the date listed. Several days before and after will be nearly as good.

Evening: Aug 20, 2010 at 46° 0' E, in Virgo, with Mars and Saturn

Morning: Jan 8, 2011 at 47° 0' W, in Libra

Evening: Mar 27, 2012 at 46° 0' E, in Aries, with the Moon and Jupiter

Morning: Aug 15, 2012 at 45° 48' W, in Gemini

Evening: Nov 1, 2013 at 47° 6' E, on the Sagittarius/Ophiuchus cusp

Morning: Mar 22, 2014 at 46° 36' W, in Capricorn

Evening: June 6, 2015 at 45° 24' E, in Cancer, with Mercury below

Morning: Oct 26, 2015 at 46° 24' W, in Leo, with Jupiter and Mars

Evening: Jan 12, 2017 at 47° 6' E, in Aquarius, with Neptune (telescope) and Mars

Morning: June 3, 2017 at 45° 54' W, in Pisces

Evening: Aug 17, 2018 at 45° 54' E, in Virgo

Morning: Jan 6, 2019 at 47° 0' W, in Libra, with Jupiter

Evening: Mar 24, 2020 at 46° 6' E, in Aries

Morning: Aug 13, 2020 at 45° 48' W, in Gemini

Evening: Oct 29, 2021 at 47° 0' E, in Scorpio/Ophiuchus

Morning: Mar 20, 2022 at 46° 36' W, in Capricorn, with Mars and Saturn

Evening: June 4, 2023 at 45° 24' E, in Cancer, with Mars

Morning: Oct 23, 2023 at 46° 24' W, in Leo

Evening: Jan 10, 2025 at 47° 12' E, in Aquarius, with Saturn

Morning: June 1, 2025 at 45° 54' W, in Pisces

Evening: Aug 15, 2026 at 45° 54' E, in Virgo, with New Crescent Moon

Morning: Jan 3, 2027 at 47° 0' W, in Libra, with Waning Crescent Moon

Evening: Mar 22, 2028 at 46° 6' E, in Aries

Morning: Aug 10, 2028 at 45° 0' W, in Gemini

Evening: Oct 27, 2029 at 47° 0' E, in Scorpio/Ophiuchus

Morning: Mar 17, 2030 at 46° 36' W, in Capricorn

Evening: June 2, 2031 at 45° 24' E, on the Gemini/Cancer cusp

Morning: Oct 21, 2031 at 46° 24' W, in Leo

Evening: Jan 7, 2033 at 47° 12' E, in Aquarius

Morning: May 29, 2033 at 45° 54' W, in Pisces

Evening: Aug 12, 2034 at 45° 54' E, in Virgo

Morning: Jan 1, 2035 at 46° 54' W, in Libra, with Mars

Evening: Mar 20, 2036 at 46° 12' E, in Aries, with Jupiter and Mars

Morning: Aug 8, 2036 at 45° 0' W, on the Taurus/Gemini cusp, with Jupiter

Evening: Oct 25, 2037 at 47° 0' E, in Scorpio/Ophiuchus

Morning: Mar 15, 2038 at 46° 36' W, in Capricorn

Evening: May 30, 2039 at 45° 24' E, in Gemini

Morning: Oct 19, 2039 at 46° 24' W, in Leo, with Jupiter

Evening: Jan 5, 2041 at 47° 12' E, in Aquarius, with New Crescent Moon

Morning: May 27, 2041 at 45° 54' W, in Pisces, with Mars and Waning Crescent Moon

Evening: Aug 10, 2042 at 45° 0' E, in Virgo, with Mars

Morning: Dec 29, 2042 at 46° 54' W, in Libra, with Saturn

Evening: Mar 17, 2044 at 46° 12' E, in Aries

Morning: Aug 6, 2044 at 45° 0' W, on the Taurus/Gemini cusp

Evening: Oct 22, 2045 at 46° 54' E, in Scorpio/Ophiuchus, with Saturn

Morning: Mar 13, 2046 at 46° 42' W, in Capricorn

Evening: May 28, 2047 at 45° 24' E, in Gemini, with New Crescent Moon

Morning: Oct 16, 2047 at 46° 18' W, in Leo, with Mars and the Waning Crescent Moon

Evening: Jan 2, 2049 at 47° 12' E, in Aquarius

Morning: May 24, 2049 at 45° 54' W, in Pisces

Evening: Aug 7, 2050 at 45° 48' E, in Virgo

Morning: Dec 27, 2050 at 46° 54' W, in Libra

Appendix C

Mars' Location and Motion in the Night Sky 2010–2050

Here are the positions and motions of wild and wayward Mars in the night sky relative to the Sun, the Moon, and the other planets, as seen through its oppositions, its larger cycle, and its retrogrades.

Sun-Mars Opposition

Only the sky position is given here, not the astrological one, but the dates are the same. Since Mars is directly opposite the Sun, it rises near sunset and sets around sunrise, moving east to west, visible through the night. Calculate solar noon for your own location—halfway between sunrise and sunset—and Mars will be overhead on the north-south meridian twelve hours later.

Jan 30, 2010	Mars in Cancer
Mar 4, 2012	Mars in Leo (25 mo, 5 days after the previous opposition)
Apr 9, 2014	Mars in Virgo (25 mo, 5 days)
May 21, 2016	Mars in Scorpio (25 mo, 12 days)
July 28, 2018	Mars in Capricorn (26 mo, 7 days)

Oct 14, 2020	Mars in Pisces (26 mo, 17 days)
Dec 7, 2022	Mars in Taurus (25 mo, 24 days)
Jan 16, 2025	Mars in Gemini (25 mo, 9 days)
Feb 20, 2027	Mars in Leo (25 mo, 4 days)
Mar 25, 2029	Mars in Virgo (25 mo, 5 days)
May 4, 2031	Mars in Libra (25 mo, 10 days)
June 27, 2033	Mars in Sagittarius (25 mo, 23 days)
Sept 17, 2035	Mars in Aquarius (26 mo, 21 days)
Nov 18, 2037	Mars in Taurus (26 mo, 1 day)
Jan 2, 2040	Mars in Gemini (25 mo, 15 days)
Feb 6, 2042	Mars in Leo (25 mo, 4 days)
Mar 12, 2044	Mars in Leo (25 mo, 6 days)
Apr 17, 2046	Mars in Virgo (25 mo, 5 days)
June 3, 2048	Mars in Scorpio (25 mo, 17 days)
Aug 15, 2050	Mars in Capricorn (26 mo, 12 days)

Mars' Cyclical Pattern

Mars' recurring pattern is best seen in its conjunctions with the Sun, but it only appears over a *long* stretch of years. Go forward seventy-nine years, plus a couple of days, and you'll find Mars, just a couple of degrees farther along from its earlier position. Some examples:

Feb 4, 2011— Aquarius 15°30'	+79 years, 4 days, 5°	Feb 8, 2090— Aquarius 20°01'
Apr 18, 2013— Aries 28° 8'	+79 years, 3 days, 4°	Apr 21, 2092— Taurus 2° 11'

June 14, 2015— Gemini 23°17'	+79 years, 3 days, 3°	June 17, 2094— Gemini 26°39'
July 27, 2017— Leo 4°12'	+79 years, 2 days, 3°	July 29, 2096— Leo 7°12'
Sept 2, 2019— Virgo 9°41'	+79 years, 2 days, 3°	Sept 4, 2098— Virgo 12°31'

Mars Retrograde 1995–2050

To work with the retrograde energies of Mars, check these dates and durations. Since retrograde work can involve looking back into your past, dates in the past are included. The "resolved" date is when Mars returns to the position it was in when that retrograde began, after which it moves into new space instead of just retracing its steps. These are given as astrological positions and, in ( ) brackets, astronomical sky positions.

Mars Goes Retrograde	Mars Goes Direct	Mars Resolves
Jan 2, 1995, Virgo 2° (Leo)	Mar 24, 1995, Leo 13° (Cancer)	May 31, 1995
Feb 6, 1997, Libra 5° (Virgo)	Apr 27, 1997, Virgo 16° (Leo)	July 1, 1997
Mar 18, 1999, Scorpio 12° (Libra)	June 4, 1999, Libra 24° (Virgo)	Aug 3, 1999
May 11, 2001, Sagittarius 29° (Sagittarius)	June 19, 2001, Sagittarius 15° (Ophiuchus)	Sept 7, 2001
July 29, 2003, Pisces 10° (Aquarius)	Sept 27, 2003, Pisces 0° (Aquarius)	Nov 8, 2003
Oct 1, 2005, Taurus 23° (Taurus)	Dec 10, 2005, Taurus 8° (Aries)	Feb 4, 2006
Nov 15, 2007, Cancer 12° (Gemini)	Jan 8, 2008, Gemini 24° (Taurus)	Apr 5, 2008

Dec 20, 2009, Leo 19° (Leo)	Mar 10, 2010, Leo 0° (Cancer)	May 17, 2010
Jan 24, 2012, Virgo 23° (Virgo)	Apr 14, 2012, Virgo 3° (Leo)	June 21, 2012
Mar 1, 2014, Libra 27° (Virgo)	May 20, 2014, Libra 9° (Virgo)	July 21, 2014
Apr 17, 2016, Sagittarius 8° (Ophiuchus)	June 29, 2016, Scorpio 23° (Libra)	Aug 22, 2016
Jun 26, 2018, Aquarius 9° (Capricorn)	Aug 27, 2018, Capricorn 28° (Capricorn)	Oct 9, 2018
Sept 10, 2020, Aries 28° (Pisces)	Nov 14, 2020, Aries 15° (Pisces)	Jan 3, 2021
Oct 30, 2022, Gemini 25° (Taurus)	Jan 12, 2023, Gemini 8° (Taurus)	Mar 16, 2023
Dec 2, 2024, Leo 6° (Cancer)	Feb 24, 2025, Cancer 17° (Gemini)	May 2, 2025
Jan 10, 2027, Virgo 10° (Leo)	Apr 1, 2027, Leo 20° (Leo)	June 8, 2027
Feb 14, 2029, Libra 13° (Virgo)	May 5, 2029, Virgo 24° (Virgo)	July 8, 2029
Mar 29, 2031, Scorpio 21° (Libra)	June 13, 2031, Scorpio 4° (Virgo)	Aug 9, 2031
May 26, 2033, Capricorn 12° (Sagittarius)	Aug 1, 2033, Sagittarius 29° (Sagittarius)	Sept 16, 2033
Aug 15, 2035, Pisces 28° (Pisces)	Oct 15, 2035, Pisces 17° (Aquarius)	Nov 28, 2035
Oct 12, 2037, Gemini 6° (Taurus)	Dec 23, 2037, Taurus 20° (Aries)	Feb 20, 2038

Nov 23, 2039, Cancer 21° (Gemini)	Feb 9, 2040, Cancer 3° (Gemini)	Apr 15, 2040
Dec 28, 2041, Leo 27° (Leo)	Mar 18, 2042, Leo 8° (Cancer)	May 25, 2042
Jan 31, 2044, Libra 0° (Virgo)	Apr 21, 2044, Virgo 11° (Leo)	June 26, 2044
Mar 11, 2046, Scorpio 6° (Virgo)	May 28, 2046, Libra 18° (Virgo)	July 28, 2046
Apr 30, 2048, Sagittarius 20° (Ophiuchus)	July 10, 2048, Sagittarius 5° (Scorpio)	Sept 1, 2048
July 14, 2050, Aquarius 26° (Aquarius)	Sept 13, 2050, Aquarius 16° (Capricorn)	Oct 23, 2050

Appendix D

Jupiter's Location and Motion in the Night Sky 2010–2050

These are the dates when Jupiter is at or near solar opposition and at its brightest for the year, receiving the Sun's light most directly. Date given is for the exact opposition, but viewing is dramatic for weeks before and after, as Jupiter is closest to Earth at these points and moving slowly in retrograde motion. To find the recurring pattern, add twelve years to any date, then add several days and about 5°. On these dates, Jupiter has a midnight meridian. Jupiter's astronomical position is given first, followed by its astrological position.

Sept 21, 2010	Pisces (Pisces 28°23')
Oct 29, 2011	Aries (Taurus 5°17')
Dec 3, 2012	Taurus (Gemini 11°17')
Jan 5, 2014	Gemini (Cancer 15°27')
Feb 6, 2015	Cancer (Leo 17°38')
Mar 8, 2016	Leo (Virgo 18°18')
Apr 7, 2017	Virgo (Libra 18°15')
May 9, 2018	Libra (Scorpio 18°21')

June 10, 2019	Ophiuchus (Sagittarius 19°28')
July 14, 2020	Sagittarius (Capricorn 22°20')
Aug 20, 2021	Capricorn/Aquarius cusp (Aquarius 27°13')
Sept 26, 2022	Pisces (Aries 3°41')
Nov 3, 2023	Aries (Taurus 10°30')
Dec 7, 2024	Taurus (Gemini 16°15')
Jan 10, 2026	Gemini (Cancer 20°06')
Feb 11, 2027	Leo (Leo 22°02')
Mar 12, 2028	Leo (Virgo 22°35')
Apr 12, 2029	Virgo (Libra 22°32')
May 13, 2030	Libra (Scorpio 22°46')
June 15, 2031	Ophiuchus (Sagittarius 24°06')
July 19, 2032	Sagittarius (Capricorn 27°14')
Aug 25, 2033	Aquarius (Pisces 2°21')
Oct 2, 2034	Pisces (Aries 8°55')
Nov 8, 2035	Aries (Taurus 15°38')
Dec 12, 2036	Taurus (Gemini 21°09')
Jan 14, 2038	Gemini (Cancer 24°44')
Feb 15, 2039	Leo (Leo 26°28')
Mar 16, 2040	Virgo (Virgo 26°55')
Apr 16, 2041	Virgo (Libra 26°52')
May 17, 2042	Libra (Scorpio 27°12')
June 20, 2043	Sagittarius (Sagittarius 28°42')

July 24, 2044	Capricorn (Aquarius 2°15')
Aug 30, 2045	Aquarius (Pisces 7°19')
Oct 7, 2046	Pisces (Aries 13°57')
Nov 13, 2047	Aries (Taurus 20°36')
Dec 17, 2048	Taurus (Gemini 25°57')
Jan 19, 2050	Gemini/Cancer cusp (Cancer 29°22')

Jupiter Relative to Venus

The dates given are for their closest conjunction, but they'll still look impressive a few days earlier or later. They tend to step forward through the zodiac in an orderly sign-by-sign manner. These are sky positions, not astrological.

Feb 16, 2010	After sunset, looking west, in Aquarius
May 11, 2011	Before dawn, looking east, in Pisces
Mar 12–13, 2012	After sunset, looking west, in Aries
June 17, 2012	Before dawn, looking east, with Venus, Jupiter, and the Waning Crescent Moon, in Taurus; a triple goddess gathering: Athene, Aphrodite, and white-armed Hera
July 15, 2012	Before dawn, looking east, in Taurus; another triple goddess gathering
May 27–28, 2013	After sunset, looking west-northwest, in Taurus
Aug 18, 2014	Before dawn, looking east, in Cancer
June 30, 2015	After sunset, looking west, in Leo
Oct 25, 2015	Before dawn, looking east-southeast, in Leo
Aug 27, 2016	After sunset, looking west, in Virgo

Nov 13, 2017	Before dawn, looking east, on the Virgo/Libra cusp
Sept 19–Oct 5, 2018	After sunset, looking west, straddling Virgo and Libra
Jan 22, 2019	Before dawn, looking east, in Ophiuchus/Scorpio
Nov 23–24, 2019	After sunset, looking southwest, in Sagittarius; a triple goddess gathering
Feb 11, 2021	Before dawn, looking east, in Capricorn
Apr 30, 2022	Before dawn, looking east, on the Aquarius/Pisces cusp
Mar 1, 2023	After sunset, looking west, in Pisces
May 23, 2024	Before dawn, looking east, in Taurus
Aug 11–12, 2025	Before dawn, looking east, in Gemini
June 9, 2026	After sunset, looking west, in Gemini
Aug 25, 2027	After sunset, looking west, in Leo
Nov 9–10, 2028	Before dawn, looking east, in Virgo
Sept 6–7, 2029	After sunset, looking southwest, in Virgo
Nov 20, 2030	After sunset, looking west, in Scorpio
Feb 6–7, 2032	Before dawn, looking east, in Sagittarius
Dec 8, 2032	After sunset, looking southwest, in Capricorn
Feb 22, 2034	After sunset, looking southwest, in Aquarius
May 17, 2035	Before dawn, looking east, in Aries
March 21–23, 2036	After sunset, looking southwest, in Aries; first of three conjunctions over next 4 months
June 4, 2036	Before dawn, looking east, in Taurus; second of three conjunctions over 4 months

July 20–27, 2036	Before dawn, looking east, in Taurus; last of three conjunctions; a triple goddess gathering
May 29–June 6, 2037	After sunset, looking west-northwest, in Gemini
Aug 22–26, 2038	Before dawn, looking east-northeast, in Cancer
July 9–13, 2039	After sunset, looking west, in Leo
Nov 2, 2039	Before dawn, looking east, in Virgo
Sept 1, 2040	After sunset, looking west, in Virgo. A triple goddess gathering
Nov 17, 2041	Before dawn, looking southeast, in Libra
Jan 28–29, 2043	Before dawn, looking southeast, in Ophiuchus/Scorpio
Nov 29, 2043	After sunset, looking southwest, in Sagittarius
Feb 16–17, 2045	Before dawn, looking southeast, in Capricorn
May 7, 2046	After sunset, looking west, in Pisces
May 28, 2048	Not visible (the Sun occults Venus on this day), in Taurus
Aug 17–18, 2049	Before dawn, looking east, in Gemini
June 14–15, 2050	After sunset, looking west, in Cancer

Jupiter Retrograde 1999–2050

These are given in astrological terms, but the dates remain the same. You can use them for doing the Jupiter Retrograde tarot spread.

Start (Astrological)	Goes Direct At	Release	Sky Positions (Astronomical)
Aug 25, 1999, Taurus 4°	Dec 20, 1999, Aries 25°	Mar 12, 2000	Aries/Pisces/Aries
Sept 29, 2000, Gemini 11°	Jan 25, 2001, Gemini 1°	Apr 20, 2001	Taurus
Nov 2, 2001, Cancer 15°	Mar 2, 2002, Cancer 5°	May 23, 2002	Gemini
Dec 4, 2002, Leo 18°	Apr 4, 2003, Leo 8°	July 2, 2003	Leo/Cancer/Leo
Jan 2, 2004, Virgo 18°	May 5, 2004, Virgo 8°	July 29, 2004	Leo
Feb 2, 2005, Libra 18°	June 5, 2005, Libra 8°	Aug 30, 2005	Virgo
Mar 4, 2006, Scorpio 18°	July 6, 2006, Scorpio 8°	Sept 29, 2006	Libra
Apr 6, 2007, Sagittarius 19°	Aug 7, 2007, Sagittarius 9°	Oct 29, 2007	Ophiuchus
May 9, 2008, Capricorn 22°	Sept 7, 2008, Capricorn 12°	Dec 1, 2008	Sagittarius
June 15, 2009, Aquarius 27°	Oct 13, 2009, Aquarius 17°	Jan 5, 2010	Aquarius/ Capricorn/Aquarius
July 23, 2010, Aries 3°	Nov 17, 2010, Pisces 23°	Feb 8, 2011	Pisces/Aquarius/Pisces

Start (Astrological)	Goes Direct At	Release	Sky Positions (Astronomical)
Aug 30, 2011, Taurus 10°	Dec 25, 2011, Taurus 0°	Mar 17, 2012	Taurus
Nov 7, 2013, Cancer 20°	Mar 6, 2014, Cancer 10°	May 31, 2014	Gemini
Dec 8, 2014, Leo 22°	Apr 8, 2015, Leo 12°	July 4, 2015	Leo
Jan 8, 2016, Virgo 23°	May 9, 2016, Virgo 13°	Aug 6, 2016	Leo
Feb 6, 2017, Libra 23°	June 9, 2017, Libra 13°	Sept 7, 2017	Virgo
Mar 9, 2018, Scorpio 23°	July 10, 2018, Scorpio 13°	Oct 6, 2018	Libra
Feb 10, 2019, Sagittarius 24°	Aug 11, 2019, Sagittarius 14°	Nov 5, 2019	Ophiuchus
May 14, 2020, Capricorn 27°	Sept 13, 2020, Capricorn 17°	Dec 6, 2020	Sagittarius
June 20, 2021, Pisces 2°	Oct 18, 2021, Aquarius 22°	Jan 9, 2022	Aquarius/ Capricorn/Aquarius
July 28, 2022, Aries 8°	Nov 23, 2022, Pisces 28°	Feb 11, 2023	Pisces
Sept 4, 2023, Taurus 15°	Dec 31, 2023, Taurus 5°	Mar 21, 2024	Aries
Oct 9, 2024, Gemini 21°	Feb 4, 2025, Gemini 11°	Apr 30, 2025	Taurus
Nov 11, 2025, Cancer 25°	Mar 11, 2026, Cancer 15°	June 6, 2026	Gemini

Start (Astrological)	Goes Direct At	Release	Sky Positions (Astronomical)
Dec 13, 2026, Leo 27°	Apr 13, 2027, Leo 17°	July 12, 2027	Leo/Cancer/Leo
Jan 12, 2028, Virgo 27°	May 12, 2028, Virgo 17°	Aug 9, 2028	Virgo/Leo/Virgo
Feb 10, 2029, Libra 27°	June 13, 2029, Libra 17°	Sept 10, 2029	Virgo
Mar 13, 2030, Scorpio 27°	July 15, 2030, Scorpio 17°	Oct 8, 2030	Libra
Apr 15, 2031, Sagittarius 28°	Aug 16, 2031, Sagittarius 19°	Nov 6, 2031	Sagittarius/ Ophiuchus/ Sagittarius
May 19, 2032, Aquarius 2°	Sept 17, 2032, Capricorn 22°	Dec 11, 2032	Capricorn/ Sagittarius/ Capricorn
June 25, 2033, Pisces 7°	Oct 23, 2033, Aquarius 27°	Jan 13, 2034	Aquarius/ Capricorn/ Aquarius
Aug 3, 2034, Aries 13°	Nov 29, 2034, Aries 4°	Feb 15, 2035	Pisces/Cetus/Pisces
Sept 9, 2035, Taurus 20°	Jan 5, 2036, Taurus 10°	Mar 26, 2036	Aries
Oct 14, 2036, Gemini 26°	Feb 9, 2037, Gemini 13°	May 5, 2037	Taurus
Nov 16, 2037, Cancer 29°	Mar 15, 2038, Cancer 19°	June 8, 2038	Cancer/Gemini/ Cancer
Dec 17, 2038, Virgo 1°	Apr 17, 2039, Leo 21°	July 14, 2039	Leo
Jan 16, 2040, Libra 1°	May 18, 2040, Virgo 21°	Aug 13, 2040	Virgo/Leo/Virgo

Start (Astrological)	Goes Direct At	Release	Sky Positions (Astronomical)
Feb 14, 2041, Scorpio 1°	June 18, 2041, Libra 21°	Sept 16, 2041	Virgo
Mar 18, 2042, Sagittarius 2°	July 19, 2042, Scorpio 22°	Oct 16, 2042	Libra
Apr 20, 2043, Capricorn 3°	Aug 20, 2043, Sagittarius 23°	Nov 12, 2043	Sagittarius/ Ophiuchus/ Sagittarius
May 24, 2044, Aquarius 6°	Sept 22, 2044, Capricorn 27°	Dec 16, 2044	Capricorn/ Sagittarius/ Capricorn
July 1, 2045, Pisces 12°	Oct 28, 2045, Pisces 2°	Jan 18, 2046	Aquarius
Aug 8, 2046, Aries 19°	Dec 4, 2046, Aries 9°	Feb 25, 2047	Pisces/Cetus/Pisces
Sept 14, 2047, Taurus 25°	Jan 10, 2048, Taurus 15°	Mar 31, 2048	Taurus/Aries/Taurus
Oct 18, 2048, Cancer 1°	Feb 14, 2049, Gemini 20°	May 22, 2049	Gemini/Taurus/ Gemini
Nov 20, 2049, Leo 4°	Mar 20, 2050, Cancer 24°	June 15, 2050	Cancer/Gemini/ Cancer

Appendix E

Saturn/Jupiter Conjunctions

These are the relatively rare dates when the two slow-moving planets Saturn and Jupiter meet in the sky. The dates given are for their closest, most exact conjunctions, but also watch a week or so before and after these dates to see them draw together and then move apart.

May 28, 2000	Aries/Taurus cusp (dawn sky, only 15° from the Sun, poor visibility)
Dec 20, 2020	Sagittarius/Capricorn cusp (evening sky, great visibility)
Oct 31, 2040	Virgo (dawn sky, 20° from the Sun, decent visibility)
Apr 7, 2060	Taurus (evening sky, great visibility)

Notes

Chapter 0: Following the North Star

1. Robert Graves, *The Greek Myths* (London: Penguin, 1980), Book I, #22.
2. The population of Athens circa 430 BCE was perhaps half a million people.
3. Catherine Tennant, *Box of Stars* (Boston: Little, Brown, 1993), p. 18.
4. Elizabeth Barber, *Women's Work* (Boston: W. W. Norton & Co., 1995), pp. 42–45.
5. A. K. Andrén, et al., *Old Norse Religion in Long-Term Perspectives* (Grand Rapids: Nordic Academic Press, 2007), p. 164.
6. Diana Paxson, *Taking Up the Runes* (New York: Weiser, 2005), p. 173.
7. Susan Gitlin-Emmer, *Lady of the Northern Light* (Freedom, CA: Crossing, 1993), pp. 78–81.
8. Bruce Dickins, *Runic and Heroic Poems of the Old Teutonic Peoples* (London: Cambridge University Press, 1915), pp. 18–19.
9. Claire R. Farrer, *Living Life's Circle* (Albuquerque: University of New Mexico Press, 1991), p. 48.

10. We see the Dippers rotate counterclockwise because we're below; but if a spun strand of yarn were descending to Earth from the Dipper, and we were spinning our own cord to match that one, both would be clockwise from our overhead perspective as the spinner.
11. Farrer, *Living Life's Circle*, p. 94.
12. Angela Davis, "Black Women and Music: A Historic Legacy of Struggle," in *Black Feminist Cultural Criticism*, Jacqueline Bobo, ed., p. 221.
13. J. Frank Dobie, *Follow de Drinkin' Gou'd* (Austin, TX: Texas Folk-lore Society, 1928), pp. 81–84.
14. Jacqueline Tobin and Raymond Dobard, *Hidden in Plain View* (New York: Doubleday, 1999).
15. E. C. Krupp, *Echoes of the Ancient Skies* (New York: Harper & Row, 1983), pp. 138–140.
16. Teng Shu-P'Ing, "The Original Significance of *Bi* Disks," *Journal of East Asian Archaeology* 2.1–2 (2000): pp. 165–194.
17. Homer, *The Odyssey*, chapter V.
18. Gavin White, *Babylonian Star-Lore* (London: Solaria Publications, 2008), p. 223.
19. Oswald Wirth, *The Tarot of the Magicians* (Boston: Houghton Mifflin, 1984), pp. 46–47; Papus, *The Tarot of the Bohemians* (New York: Samuel Weiser, 1971), p. 249.
20. Why midnight? Because local sunset and twilight times vary so widely that there's no telling when the sky would actually be dark enough to allow viewing. If you're watching before midnight, spin the picture back in the *opposite* direction of the arrows.
21. All sky images are based on *Starry Night Pro* astronomical software, unless stated otherwise.
22. White, *Babylonian Star-Lore*, p. 220.
23. David Fitzgerald, "Myths of the Stars, Light and Time," *The Gentleman's Magazine*, CCLVIII (1885), pp. 507–508.

Chapter 1: The Zodiac—Our Circle of Animals

24. Clifford Pickover, *A Passion for Mathematics* (Hoboken, NJ: Wiley, 2005), p. 270.
25. Anthony Aveni, *The End of Time: The Maya Mystery of 2012* (Boulder, CO: University Press of Colorado, 2009), p. 103.
26. Veronica Ions, *Egyptian Mythology* (New York: Peter Bedrick, 1982), p. 92 passim.
27. Richard Hinckley Allen, *Star Names: Their Lore and Meaning* (New York: Dover Publications, 1963), p. 385.
28. Buffie Johnson, *Lady of the Beasts* (San Francisco: HarperSanFrancisco, 1990), pp. 102–103.
29. Ean Begg, *The Cult of the Black Virgin* (London: Ardana/Penquin, 1985), pp. 56–61.
30. White, *Babylonian Star-Lore*, pp. 112–118.
31. Allen, *Star Names,* p. 275.
32. Allen, *Star Names,* p. 366.
33. White, *Babylonian Star-Lore*, p. 157.
34. Robert Graves, *The Greek Myths*; Jo Forty, *Classic Mythology* (San Diego, CA: Thunder Bay, 1999), p. 58; Gavin White, *Babylonian Star-Lore*, pp. 155–161; Lindsay River and Sally Gillespie, *The Knot of Time* (New York: Harper & Row, 1989), p. 188.
35. John N. Wilford, "Ruins May Yield Clues on B.C. Medicine," *The Denver Post,* June 21, 1990.
36. White, *Babylonian Star-Lore,* pp. 48–49.
37. Al-Biruni, *Book of Instructions in the Elements of the Art of Astrology* (New York: Astrology Classics, 2006), p. 45.

Chapter 2: The Dance of the Sun

38. Daily motion and positions are gauged from Pottenger, *The New American Ephemeris for the 21st Century* (Grand Rapids, MI: Starcrafts LLC, 2006).
39. Dirk Jan Struik, *A Concise History of Mathematics* (New York: Dover Publications, 1987), p. 118.
40. Janet Farrar, *The Witches' God* (Custer, WA: Phoenix, 1989), p. 319; Robert Graves, *The White Goddess,* (New York: Noonday, 1966), p. 283; Barbara G. Walker, *The Women's Encyclopedia* (San Francisco: Harper & Row, 1983), p. 166.
41. Walker, *The Women's Encyclopedia,* p. 1098.
42. Anthony Murphy and Richard Moore, *Island of the Setting Sun* (Dublin: Liffey, 2008), p. 173.
43. Sheena McGrath, *The Sun Goddess* (New York: Blandford, 1997), p. 20.
44. Walker, *The Women's Encyclopedia,* p. 167.
45. Christopher Knight and Robert Lomas, *Uriel's Machine* (Gloucester, MA: Fair Winds, 2001), p. 173.
46. Susan Gitlin-Emmer, *Lady of the Northern Light* (Freedom, CA: Crossing, 1993), pp. 73–77.
47. I learned of Rösaring while searching for far-northern sites. Information about this under-studied site is drawn from articles by Börje Sandén, Emilia Pásztor, and others (listed by author in the bibliography) and from a talk given by Christine Foltzer at the East Coast Thing 2009.
48. Madhusree Mukerjee, "Circles for Space," *Scientific American* (December 8, 2003).
49. Gabriel Cooney, et al., *Brú na Bóinne* (Wicklow: Archaeology Ireland, 2003), pp. 12–14.
50. *The Mystery of Chaco Canyon,* directed by Anna Sofaer (Oley, PA: Bullfrog Films), 1999; Anna Sofaer, *Chaco Astronomy* (Santa Fe, NM: Ocean Tree, 2007), pp. 23–37.

51. Andreas Volwahsen, *Cosmic Architecture in India* (Munich: Prestel, 2001), pp. 70–71.

Chapter 3: The Precession of the Equinoxes

52. "Pisces 8°" or "8° of Pisces" are astrological expressions of distance, based on each sign having 30° of space, which isn't physically accurate.
53. The exact duration is currently interpreted as 25,770.1 years, per: Anthony Aveni, *The End of Time: The Maya Mystery of 2012* (Boulder, CO: University Press of Colorado, 2009), p. 100.
54. Patrick Symmes, "History in the Remaking," *Newsweek*, March 1, 2010, pp. 46–48; Andrew Curry, "Gobekli Tepe: The World's First Temple?" *Smithsonian*, November 2008.
55. Allen, *Star Names*, p. 206.
56. John Barrow, *The Artful Universe* (New York: Oxford University Press, 1995), p. 141; E. C. Krupp, *Echoes of the Ancient Skies* (New York: Harper & Row, 1983), pp. 100 passim; Anthony Aveni, *People and the Sky* (New York: Thames and Hudson, 2008), p. 166.
57. Masaru Emoto, *Messages from Water* (Netherlands: Hado Publishing, 1999) and other works.
58. Sandra Ingerman, *Medicine for the Earth* (New York: Three Rivers, 2000) and other works.

Chapter 4: The Moon—Queen of the Night

59. Anne Kent Rush, *Moon, Moon* (New York: Random House, 1976), pp. 300–304.
60. Positions and declination based on *Starry Night Pro* software.
61. These maximum declinations, north and south, are commonly rounded to 28.5° for ease of dealing with them.

62. This is from March 2006.

63. These minimum declinations, north and south, are commonly rounded to 18.5° for ease of dealing with them.

64. This is from March 2015.

65. *The Mystery of Chaco Canyon.*

66. Anna Sofaer, speaking in *The Mystery of Chaco Canyon.*

67. E. G. Squier and E. H. Davis, *Ancient Monuments of the Mississippi Valley* (Washington, DC: Smithsonian Institution, 1998. Reprint from 1848), Plate XXV.

68. Ohio's Serpent Mound is also purported to have lunar extreme sightlines, aimed along the body's coils, but some writers claim the same coils and sightlines as solstice/equinox alignments. They can't be both, might be neither, and hopefully will attract more research.

69. Emilia Pásztor, et al. "The Sun and the Rösaring Ceremonial Road." *European Journal of Archaeology*, April (2000).

70. A compromise between the sidereal month's 27.3 days and the synodic month's 29.53 days? More likely, this comes from the 28 days of actual visibility of each cycle, not counting the unseen Dark Moon. In either case, this 28-day idea is deeply embedded.

Chapter 5: Mercury—Magical Messenger and Soul Guide

71. Luisah Teish, *Jambalaya* (San Francisco: Harper & Row, 1985), pp. 108–109, 113–117.

72. Samuel N. Kramer, *The Sumerians* (Chicago: University of Chicago Press, 1971), p. 138.

73. *The Old Farmer's 2010 Almanac*, p. 95. There's also a website that offers rising and setting times for all the planets. The calculator can be set for a range of dates and United States locations. See *www.almanac.com.*

74. Chalice of Repose Project, P.O. Box 169, Mt. Angel, OR, 97362; See *www.chaliceofrepose.org.*

75. Training resources, usually listed as "Death and Dying" workshops, can be found as "Advanced Workshops" at the website *www.shamanicteachers.com.*

Chapter 6: Venus—A Walk with Love, Death, and Rebirth

76. Diane Wolkstein and Samuel N. Kramer, *Inanna* (New York: Harper & Row, 1983), p. 12. All following quotes from the Inanna poetry are also from this source.
77. Carolyn Larrington, trans., *Poetic Edda* (Oxford: Oxford University Press, 1999), p. 90.
78. Jacob Grimm, *Teutonic Mythology*, Vol II (London: George Bell and Sons, 1883), p. 723.
79. Gitlin-Emmer, *Lady of the Northern Lights* (London: George Bell and Sons, 1883), p. 86–87.
80. Gerald Hawkins, *Mindsteps* (New York: Harper & Row, 1983), pp. 46–69; Christopher Knight and Robert Lomas, *Uriel's Machine*, pp. 102–103; Henry Lincoln, *Key to the Sacred Pattern: the Untold Story of Rennes-le-Château.* (New York: St. Martin's, 1998), pp. 144–145.
81. Mary K. Greer, *Tarot Constellations* (North Hollywood, CA: Newcastle Publishing, 1987), pp. 206–207.
82. Lindsay River and Sally Gillespie, *The Knot of Time* (New York: Harper & Row, 1989), p. 20.
83. Wolkstein and Kramer, *Inanna,* p. 101.
84. Wolkstein and Kramer, *Inanna,* pp. 37–48.
85. Wolkstein and Kramer, *Inanna,* p. 40.
86. Hawkins, *Mindsteps*, pp. 46–69; Anthony Aveni, *Skywatchers of Ancient Mexico* (Austin: University of Texas Press, 1980), pp. 86, 89 and note #15. Citation taken from "Nägele's Rule."
87. Walker, *The Women's Encyclopedia*, p. 682.
88. Layne Redmond, *When the Drummers Were Women* (New York: Three Rivers, 1997), p. 76.

89. Wolkstein and Kramer, *Inanna*, pp. 52–73. Earlier, less complete translations of the descent story have *Dumuzi* in the underworld and Inanna going to *his* rescue. As more tablet fragments were located, the tale changed shape, as clarified in Samuel Kramer's *History Begins at Sumer: Thirty-Nine Firsts in Recorded History* (Philadelphia: University of Pennsylvania Press, 1956), pp. 166–167.

90. Wolkstein and Kramer, *Inanna,* p. 52.

91. Wolkstein and Kramer, *Inanna,* p. 55.

92. Michael S. Schneider, *Beginner's Guide to Constructing the Universe* (New York: HarperCollins, 1994), p. 136.

93. Wolkstein and Kramer, *Inanna,* p. 60.

94. Wolkstein and Kramer, *Inanna,* p. 71.

95. Wolkstein and Kramer, *Inanna,* pp. 71–76.

96. Knight and Lomas, *Uriel's Machine*, pp. 276–285; Murphy and Moore, *Island of the Setting Sun*, pp. 164–168. The books differ on which year of the Venus cycle is the intended focus.

97. Venus is a Morning Star half the time, with five distinct morning-sky appearances, but the idea is that the light box is aimed at a *particular* cycle of Morning Star appearances.

98. Murphy and Moore, *Island of the Setting Sun*, p. 164.

99. Knight and Lomas, *Uriel's Machine*, p. 269.

100. Eli Maor, *June 8, 2004—Venus in Transit* (Princeton, NJ: Princeton University Press, 2000), p. 173.

Chapter 7: Mars—A Planetary Rebel?

101. J. L. E. Dreyer, *A History of Astronomy* (New York: Dover Publications, 1953), p. 101.

102. John Neihardt, *Black Elk Speaks* (New York: Pocket, 1977), pp. 46–47.

103. Ralph Waldo Emerson, *Emerson: Essays and Lectures* (New York: Library of America, 1983), from "An Introductory Lecture on the Times," p. 166.

104. Lord Byron, *Don Juan* (New York: Modern Library, 1949), Canto the Thirteenth, CI.

105. Richard Louv, *Last Child in the Woods* (Chapel Hill, NC: Algonquin of Chapel Hill, 2005), pp. 98–111.

Chapter 8: Jupiter—King of Many Names

106. Leonard W. King, *The Seven Tablets of Creation* (London: FQ Classics, 2007).

107. White, *Babylonian Star-Lore*, p. 281.

108. Herbert Bates, trans., *The Odyssey of Homer* (New York: Harper and Brothers, 1929), p.183.

109. Florence Wood and Kenneth Wood, *Homer's Secret Iliad: The Epic of the Night Skies Decoded* (New York: John Murray, 1999).

110. Jacob Grimm, *Teutonic Mythology.* Trans. Stallybras (London: George Bell and Sons, 1883), vol. IV, p. 17–37.

111. Peg Weiss, *Kandinsly and Old Russia* (New Haven: Yale University Press, 1995), pp. 84–85.

112. Nigel Pennick, *Magical Alphabets* (York Beach, ME: Samuel Weiser, 1992), pp. 136–137.

113. For example, on August 25, 1999, Jupiter went retrograde at 4° Taurus. It moved slowly backward to 12° Aquarius before going direct on October 5, and finally "released"—returned to 4° Taurus—in mid-March 2000. Twelve years and five days later, on August 30, 2011, Jupiter went retrograde at 10° Taurus. It moved backward to 25° Aries, then went direct on December 20, and released at 10° Taurus in mid-March 2012. The 2011–2012 Jupiter retrograde period is used as a lens to reflect back on life during the 1999–2000 retrograde.

Chapter 9: Saturn—The Ancients' Final Frontier

114. Uranus was discovered in 1781, Neptune in 1846. Pluto was discovered in 1930 but demoted in 2006 from planet to "dwarf planet," a sort of demi-planet sky object.

115. Giorgio de Santillana and Hertha von Dechend, *Hamlet's Mill* (San Diego: David R. Godine, 1992), pp. 135–136.

116. de Santillana and von Dechend, *Hamlet's Mill,* p. 129.

117. de Santillana and von Dechend, *Hamlet's Mill,* p. 129, n. 45.

118. Graves, *The Greek Myths*, Book 1, pp. 39–44.

119. "Five Wishes" forms come from Aging with Dignity, PO Box 1661, Tallahassee, Florida 32302-1661, *www.agedwithdignity.org.*

Chapter 10: Some Special Stars, Groups, and Phenomena

120. John Eddy, "Probing the Mystery of the Medicine Wheels," *National Geographic,* 151.1 (1977): 140–46.

121. Nev E. Autrey and Wanda R. Autrey, "Zodiac Ridge," in *Archaeoastronomy in the Americas*, Williamson (ed.), pp. 81–99.

122. Autrey, "Zodiac Ridge," p. 95.

123. David Freidel et al., *Maya Cosmos* (New York: Harper Paperbacks, 1995), pp. 85–100.

124. Freidel, et al., *Maya Cosmos,* p. 107.

125. Anthony Aveni, *The End of Time: The Maya Mystery of 2012* (Boulder, CO: University Press of Colorado, 2009), p. 57.

126. Louis Charpentier, *The Mysteries of Chartres Cathedral* (New York: Avon, 1975), pp. 30–32; Renna Shesso, *Math for Mystics* (San Francisco: Red Wheel/Weiser, 2007), p. 72.

127. Caroline Hall Hovey, *The Somerset Sanctuary* (Devon: Merlin, 1985).

128. Gary David, *The Orion Zone* (Kempton, IL: Adventures Unlimited, 2006).

129. Robert Graves, *The Greek Myths* (London: Penguin, 1980), Book I, pp. 151–154.

130. Freidel, et al., *Maya Cosmos*, p. 79.

131. Freidel, et al., *Maya Cosmos*, p. 79.

132. Jacob Grimm, *Teutonic Mythology* (London: George Bell and Sons, 1883), p. 97.

133. A. K. Andrén, et al., *Old Norse Religion in Long-Term Perspectives* (Grand Rapids: Nordic Academic Press, 2007), pp. 164–166.

134. Grimm, *Teutonic Mythology*, p. 97.

135. Mircea Eliade, *Shamanism: Archaic Techniques of Ecstasy* (Princeton, NJ: Princeton University Press, 1972), pp. 385–386; Grimm, *Teutonic Mythology*, p. 97.

136. John Sephton, trans. *Eirik the Red's Saga: A Translation.* (Liverpool, England: D. Marples, 1880), pp. 12–15.

137. Sandra Ingerman, *Medicine for the Earth*, p. 189 passim.

Selected Bibliography

Al-Biruni. *Book of Instructions in the Elements of the Art of Astrology*. New York: Astrology Classics, 2006.

Allen, Richard Hinckley. *Star Names*. New York: Dover Publications, 1963.

Andrén, A., K. Jennbert, and C. Raudvere. *Old Norse Religion in Long-Term Perspectives: Origins, Changes and Interactions, an International Conference in Lund, Sweden, June 3–7, 2004*. Grand Rapids: Nordic Academic Press, 2007.

Aveni, Anthony F. *Conversing with the Planets: How Science and Myth Invented the Cosmos*. New York: Times, 1992.

———. *The End of Time: The Maya Mystery of 2012*. Boulder, CO: University Press of Colorado, 2009.

———. *Native American Astronomy*. Austin: University of Texas, 1977.

———. *Skywatchers of Ancient Mexico*. Austin: University of Texas, 1980.

———. *People and the Sky: Our Ancestors and the Cosmos*. New York: Thames and Hudson, 2008.

Barber, Elizabeth Wayland. *Women's Work: The First 20,000 Years—Women, Cloth, and Society in Early Times*. Boston: W. W. Norton & Co, 1995.

Barrow, John D. *The Artful Universe*. New York: Oxford University Press, 1995.

Bates, Herbert, trans. *The Odyssey of Homer*. New York: Harper and Brothers, 1929.

Begg, Ean. *The Cult of the Black Virgin*. London: Ardana/Penguin, 1985.

Bobo, Jacqueline, ed. *Black Feminist Cultural Criticism*. Malden, MA: Blackwell, 2001.

Budge, E. A. Wallis. *The Gods of the Egyptians*. Vol. 2. New York: Dover Publications, 1969.

Butler, Alan. *The Goddess, the Grail and the Lodge: Tracing the Origins of Religion*. Winchester, England: O, 2004.

Byron, George Gordon, Lord. *Don Juan*. New York: Modern Library, 1949.

Casey, Caroline. *Making the Gods Work for You: The Astrological Language of the Psyche*. New York: Three Rivers, 1999.

Charpentier, Louis. *The Mysteries of Chartres Cathedral*. New York: Avon, 1975.

Cooney, Gabriel, et al. *Brú Na Bóinne: Newgrange, Knowth, Dowth and the River Boyne*. Bray, Co. Wicklow: Archaeology Ireland, 2003.

Cornelius, Geoffrey, and Paul Devereux. *The Secret Language of the Stars and Planets: A Visual Key to the Heavens*. San Francisco: Chronicle, 1996.

Crossley-Holland, Kevin. *Norse Myths*. New York: Pantheon, 1980.

Curry, Andrew. "Gobekli Tepe: The World's First Temple?" *Smithsonian,* November (2008).

David, Gary A. *The Orion Zone: Ancient Star Cities of the American Southwest*. Kempton, IL: Adventures Unlimited, 2006.

Davidson, Hilda Roderick Ellis. *Lost Beliefs of Northern Europe*. London: Routledge, 1993.

De Santillana, Giorgio, and Hertha Von Dechend. *Hamlet's Mill: An Essay on Myth and the Frame of Time*. San Diego: David R. Godine, 1992.

Dickins, Bruce. *Runic and Heroic Poems of the Old Teutonic Peoples*. London: Cambridge University Press, 1915.

Dobie, J. Frank, ed. *Follow De Drinkin' Gou'd.* Austin, TX: Texas Folk-lore Society, 1928.

Dreyer, J. L. E. *A History of Astronomy from Thales to Kepler.* New York: Dover Publications, 1953.

Eddy, John. "Probing the Mystery of the Medicine Wheels." *National Geographic,* 151.1 (1977): 140–146.

Eliade, Mircea. *Shamanism: Archaic Techniques of Ecstasy.* Princeton, NJ: Princeton University Press, 1972.

Emerson, Ralph Waldo. *Emerson: Essays and Lectures.* New York: Library of America, 1983.

Emoto, Masaru. *Messages from Water: The First Pictures of Frozen Water Crystals.* Netherlands: Hado Publishing, 1999.

Erman, Adolf. *Life in Ancient Egypt.* Trans. Helen Mary Tirard. London: Macmillan, 1894.

Farrar, Janet, and Stewart Farrar. *The Witches' God: Lord of the Dance.* Custer, WA: Phoenix, 1989.

Farrer, Claire R. *Living Life's Circle: Mescalero Apache Cosmovision.* Albuquerque: University of New Mexico Press, 1991.

Fitzgerald, David. "Myths of the Stars, Light and Time." *The Gentleman's Magazine,* CCLVIII (1885): 507–508.

Foltzer, Christine. "Rösaring—Cult Site." *Ravencast—The Asatru Podcast.* Ravencast, 13 Jan. 2010. Webcast of a talk given at the East Coast Thing 2009. *ravencast.podbean.com.*

Forty, Jo, ed. *Classic Mythology.* San Diego, CA: Thunder Bay, 1999.

Freidel, David, Linda Schele, and Joy Parker. *Maya Cosmos.* New York: Harper Paperbacks, 1995.

Gitlin-Emmer, Susan. *Lady of the Northern Light: A Feminist Guide to the Runes.* Freedom, CA: Crossing, 1993.

Graves, Robert. *The Greek Myths.* London: Penguin, 1980.

———. *The White Goddess: A Historical Grammar of Poetic Myth*. New York: Noonday, 1966.

Greer, Mary K. *Tarot Constellations: Patterns of Personal Destiny*. North Hollywood, CA: Newcastle Publishing, 1987.

Grimm, Jacob. *Teutonic Mythology*. Trans. James S. Stallybrass. Vol. II. London: George Bell and Sons, 1883.

Hawkins, Gerald S. *Mindsteps to the Cosmos*. New York: Harper & Row, 1983.

———. *Stonehenge Decoded*. Garden City, NY: Doubleday, 1965.

Hovey, Caroline Hall. *The Somerset Sanctuary*. Devon: Merlin, 1985.

Ingerman, Sandra. *Medicine for the Earth: How to Transform Personal and Environmental Toxins*. New York: Three Rivers, 2000.

Ions, Veronica. *Egyptian Mythology*. New York: Peter Bedrick, 1982.

Johnson, Buffie. *Lady of the Beasts: Ancient Images of the Goddess and Her Sacred Animals*. San Francisco: HarperSanFrancisco, 1990.

Kennedy, Roger G. *Hidden Cities: the Discovery and Loss of Ancient North American Civilization*. New York: Penguin, 1994.

King, Leonard W. *Enuma Elish: The Seven Tablets of Creation*. London: FQ Classics, 2007.

Knight, Christopher, and Robert Lomas. *Uriel's Machine: Uncovering the Secrets of Stonehenge, Noah's Flood, and the Dawn of Civilization*. Gloucester, MA: Fair Winds, 2001.

Kramer, Samuel N. *Cradle of Civilization*. New York: Time-Life, 1967.

———. *The Sumerians: Their History, Culture, and Character*. Chicago: University of Chicago Press, 1971.

Krupp, E. C. *Echoes of the Ancient Skies the Astronomy of Lost Civilizations*. New York: Harper & Row, 1983.

Larrington, Carolyne, trans. *The Poetic Edda*. Oxford, England: Oxford University Press, 1999.

Levy, David H. *A Guide to Skywatching*. Ed. John O'Byrne. San Francisco: Fog City, 2002.

Lincoln, Henry. *Key to the Sacred Pattern: The Untold Story of Rennes-le-Château*. New York: St. Martin's, 1998.

Lippard, Lucy R. *Overlay: Contemporary Art and the Art of Prehistory*. New York: Pantheon, 1983.

Littmann, Mark, and Donald K. Yeomans. *Comet Halley: Once in a Lifetime*. Washington, DC: American Chemical Society, 1985.

Louv, Richard. *Last Child in the Woods: Saving Our Children from Nature-Deficit Disorder*. Chapel Hill, NC: Algonquin of Chapel Hill, 2005.

Malville, J. McKim, and Claudia Putnam. *Prehistoric Astronomy in the Southwest*. Rev. ed. Boulder, CO: Johnson, 1993.

Maor, Eli. *June 8, 2004: Venus in Transit*. Princeton, NJ: Princeton University Press, 2000.

McGrath, Sheena. *Sun Goddess: Myth, Legend and History*. New York: Blandford, 1997.

Michell, John F. *The New View over Atlantis*. San Francisco: Harper & Row, 1983.

Mukerjee, Madhusree. "Circles for Space." *Scientific American* (2003).

Murphy, Anthony, and Richard Moore. *Island of the Setting Sun: In Search of Ireland's Ancient Astronomers*. Dublin: Liffey, 2008.

Murray, Alexander S., and William H. Klapp. *Manual of Mythology: Greek and Roman, Norse and Old German, Hindoo and Egyptian Mythology*. Philadelphia: Henry Altemus, 1897.

The Mystery of Chaco Canyon. Oley, PA: Bull Frog Films, 1999. DVD.

Neihardt, John G., comp. *Black Elk Speaks, Being the Life Story of a Holy Man of the Oglala Sioux*. New York: Pocket, 1977.

Old Farmer's Almanac 2011, 219th edition (2010), and other years.

Papus. *The Tarot of the Bohemians: The Most Ancient Book in the World for the Exclusive Use of Initiates*. New York: Samuel Weiser, 1971.

Pásztor, Emilia, Curt Roslund, Britt-Mari Näsström, and Heather Robertson. "The Sun and the Rösaring Ceremonial Road." *European Journal of Archaeology,* April (2000).

Paxson, Diana L. *Taking Up the Runes: A Complete Guide to Using Runes in Spells, Rituals, Divination, and Magic.* New York: Samuel Weiser, 2005.

Pennick, Nigel. *Complete Illustrated Guide to Runes.* Boston, MA: Element, 1999.

————. *Magical Alphabets.* York Beach, ME: Samuel Weiser, 1992.

————. *Runic Astrology.* Grand Rapids, MI: Capall Bann, 1995.

Pickover, Clifford A. *A Passion for Mathematics: Numbers, Puzzles, Madness, Religion, and the Quest for Reality.* Hoboken, NJ: Wiley, 2005.

Pottenger, Rique. *The New American Ephemeris for the 21st Century, 2000–2100 at Midnight.* Grand Rapids, MI: Starcrafts LLC, 2006.

Randall, E. O. *The Serpent Mound, Adams County, Ohio: Mystery of the Mound and History of the Serpent: Various Theories of the Effigy Mounds and the Mound Builders.* Columbus, OH: Ohio State Archæological and Historical Society, 1905.

Reddy, Francis, and Greg Walz-Chojnacki. *Celestial Delights: The Best Astronomical Events through 2010.* Berkeley, CA: Celestial Arts, 2002.

Redmond, Layne. *When the Drummers Were Women: A Spiritual History of Rhythm.* New York: Three Rivers, 1997.

River, Lindsay, and Sally Gillespie. *The Knot of Time: Astrology and the Female Experience.* New York: Harper & Row, 1989.

Rush, Anne Kent. *Moon, Moon.* New York: Random House, 1976.

Rydberg, Viktor. *Teutonic Mythology: Gods and Goddesses of the Northland.* Trans. Rasmus Björn Anderson. New York: Norrœna Society, 1907.

Sandén, Börje. "Fifty Years with the Cult Site of Rösaring." Trans. Heather Robertson. *Viking Heritage Magazine* (2002).

Schneider, Michael S. *A Beginner's Guide to Constructing the Universe: The Mathematical Archetypes of Nature, Art, and Science*. New York: HarperCollins, 1994.

Schroeder-Sheker, Therese. *Transitus: A Blessed Death in the Modern World.* Missoula, MT: St. Dunstan's, 2001.

Sephton, John, trans. *Eirik the Red's Saga: A Translation*. Liverpool, England: D. Marples, 1880.

Shesso, Renna. *Math for Mystics: From the Fibonacci Sequence to Luna's Labyrinth to Golden Sections and Other Secrets of Sacred Geometry*. San Francisco: Red Wheel/Weiser, 2007.

Shu-P'Ing, Teng. "The Original Significance of *Bi* Disks: Insights Based on Liangzhu Jade Bi with Incised Symbolic Motifs." *Journal of East Asian Archaeology* 2.1–2 (2000): 165–94.

Sofaer, Anna. *Chaco Astronomy: An Ancient American Cosmology*. Santa Fe, NM: Ocean Tree, 2007.

Squier, E. G., and E. H. Davis. *Ancient Monuments of the Mississippi Valley.* Washington, DC: Smithsonian Institution, 1998. Reprint from 1848.

Starry Night Pro. Version. 4.5. Toronto, Ontario, Canada: Space Software, 2003. Computer software.

Struik, Dirk Jan. *A Concise History of Mathematics*. New York: Dover Publications, 1987.

The Sun Dagger. Oley, PA: Bullfrog Films, 1983.

Symmes, Patrick. "History in the Remaking." *Newsweek,* March 1 (2010): 46–48.

Teish, Luisah. *Jambalaya: The Natural Woman's Book of Personal Charms and Practical Rituals*. San Francisco: Harper & Row, 1985.

Tennant, Catherine. *Box of Stars: A Practical Guide to the Night Sky and to Its Myths & Legends*. Boston: Little, Brown, 1993.

Thom, Alexander. *Megalithic Sites in Britain*. Oxford, England: Clarendon Press, 1967.

Thomas, N. L. *Irish Symbols of 3500 BC.* Cork: Mercier, 1988.

Tobin, Jacqueline L., and Raymond G. Dobard, PhD. *Hidden in Plain View: A Secret Story of Quilts and the Underground Railroad.* New York: Doubleday, 1999.

Volwahsen, Andreas. *Cosmic Architecture in India: The Astronomical Monuments of Maharaja Jai Singh II.* Munich: Prestel, 2001.

Waite, Arthur Edward. *The Pictorial Key to the Tarot.* Mineola, NY: Dover Publications, 2005.

Walker, Barbara G. *Woman's Encyclopedia of Myths and Secrets.* San Francisco: Harper & Row, 1983.

Weiss, Peg. *Kandinsky and Old Russia: The Artist as Ethnographer and Shaman.* New Haven: Yale University Press, 1995.

White, Gavin. *Babylonian Star-Lore: An Illustrated Guide to the Star-Lore and Constellations of Ancient Babylonia.* London: Solaria Publications, 2008.

Wilford, John N. "Ruins May Yield Clues on B.C. Medicine." *The Denver Post,* June 21, 1990.

Williamson, Ray A. *Archaeoastronomy in the Americas.* Los Altos, CA: Ballena, 1981.

Williamson, Ray A. *Living the Sky: the Cosmos of the American Indian.* Boston: Houghton Mifflin, 1984.

Wirth, Oswald. *Tarot of the Magicians.* York Beach, ME: Samuel Weiser, 1985.

Wolkstein, Diane, and Samuel Noah Kramer. *Inanna, Queen of Heaven and Earth: Her Stories and Hymns from Sumer.* New York: Harper & Row, 1983.

Wood, Florence, and Kenneth Wood. *Homer's Secret Iliad: The Epic of the Night Skies Decoded.* New York: John Murray, 1999.

Illustrations

All images, other than those noted below, were created by the author, either drawn by hand, as photographs, or created in Adobe Illustrator. They are often based on astronomical material generated in *Starry Night Pro* astronomy software.

Figures 4, 110, 111: Photographs by Deb Hoffman, used by permission.

Figure 5: Virginia McGaw, *Construction Work for Rural and Elementary Schools* (Chicago: A. Flanagan Company, 1909). Public domain.

Figure 6: Jim Harter, ed. *Women: A Pictorial Archive from Nineteenth Century Sources* (Mineola, NY: Dover Publications, 1978). Used by permission.

Figures 12 and 130: Adolph Erman, *Life in Ancient Egypt*, 1894 (reprint: New York: Dover Publications, 1971). Used by permission.

Figures 14, 44, 45, 46, 47, 100, 107, 116, 126, 136: A. E. Waite, *The Pictorial Key to the Tarot*, 1911 (reprint: Mineola, NY: Dover Publications). Used by permission.

Figure 90: Traditional design, anonymous, author's collection.

Figures 91, 92 (cropped), 121, 127, 142: Alexander S. Murray and William H. Klapp, *Manual of Mythology: Greek and Roman* (Philadelphia: Henry Altemus, 1897). Public domain.

Figure 120: A. E. Wallis Budge, *The Gods of the Egyptians*, vol. 2, 1904 (reprint: New York: Dover Publications, 1969). Used by permission.

Figure 122: Victor Rydberg, *Teutonic Mythology: Gods and Goddesses of the Northland* (New York: Norroena Society, 1906). Public domain.

Index

About the Author

Renna Shesso has been a student of mystical traditions and spiritual self-discovery since the late 1960s. Initially taught by her herbalist/astrologer grandmother, she is an independent researcher on mythology and history, art and archeology, tarot, and the vast lore of the Goddess traditions. A longtime resident of Colorado, Shesso follows her calling as shamanic healing practitioner and teacher, professional tarot reader, and priestess of the Craft. Visit her at *www.rennashesso.com.*

To Our Readers

Weiser Books, an imprint of Red Wheel/Weiser, publishes books across the entire spectrum of occult, esoteric, speculative, and New Age subjects. Our mission is to publish quality books that will make a difference in people's lives without advocating any one particular path or field of study. We value the integrity, originality, and depth of knowledge of our authors.

Our readers are our most important resource, and we appreciate your input, suggestions, and ideas about what you would like to see published.

Visit our website *www.redwheelweiser.com* where you can subscribe to our newsletters and learn about our upcoming books, exclusive offers, and free downloads.

You can also contact us at info@redwheelweiser.com or at

Red Wheel/Weiser, LLC
665 Third Street, Suite 400
San Francisco, CA 94107